The Politics and Economics of Removing Subsidies on Petroleum Products in Nigeria

Published by

Adonis & Abbey Publishers Ltd

United Kingdom
PO Box 43418
London
SE11 4XZ
Email: editor@adonis-abbey.com,
Tel: 0845 388 7248

Nigeria
No.3 Akanu Ibiam Street,
Aso-villa, Asokoro.
P.O. Box 10546
Abuja
Tel: +234 (0) 8165970458, 07066997765

Year of Publication 2014

Copyright© Jideofor Adibe

British Library Cataloguing-in-Publication Data
A catalogue record for this book is available from the British Library

ISBN: 9781906704933 (HB)/9781906704940(PB)

The Politics and Economics of Removing Subsidies on Petroleum Products in Nigeria

Edited by

Jideofor Adibe

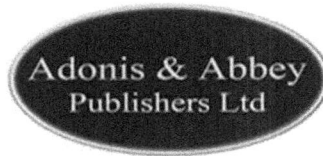

Adonis & Abbey
Publishers Ltd

TABLE OF CONTENTS

Table of Contents | Jideofor Adibe (Ed.)
The Politics and Economics of Removing Subsidies on Petroleum Products in Nigeria
London & Abuja, Adonis & Abbey Publishers

iv

Table of	Jideofor Adibe (Ed.)
contents	*The Politics and Economics of Removing Subsidies on Petroleum Products in Nigeria*
	London & Abuja, Adonis & Abbey Publishers

List of contributors

Dr Jideofor Adibe (M.Sc, Political science; Ph.D, Development Studies; LLM, Law) is a Senior Lecturer in political science at Nasarawa State University, Keffi, Nigeria. He is also the founding editor of the international academic journal *African Renaissance* which has been published continuously since June 2004. He has published several books and articles and is also on the Board of six peer-reviewed international academic journals. He is equally a columnist with the *Daily Trust* – one of Nigeria's leading national papers – and a member of the paper's Editorial Board. Dr Adibe can be reached at: pcjadibe@yahoo.com

Dr. Emmanuel Ojameruaye (Ph.D. Economics) is Vice-President for Research and Programme Development with the International Foundation for Education & Self-Help, Phoenix, United States. He held several management positions in the Shell Petroleum Development Company of Nigeria between 1992 and 2002; was a National Consultant Economist/Statistician for the National Data Bank Project from 1989 to 1992; and taught Econometrics and Statistics at the University of Benin, Benin City between 1982 and 1989. He was also the National Secretary of the Nigerian Economic Society from 1997 – 1990.

Dr Ogujiuba Kanayo (Ph.D, Economics) is a Senior Research Fellow (Public Finance/Macroeconomics Unit) at the National Institute for Legislative Studies; National Assembly, Abuja, Nigeria. He has been a consultant/senior policy adviser on World Bank, USAID, DFID, ECA, UNDP, EU, WAMI projects and to government agencies in Nigeria within the last 15 years. He is a member of several research and management

List of
Contributors | Jideofor Adibe (Ed.)
The Politics and Economics of Removing Subsidies on Petroleum Products in Nigeria
London & Abuja, Adonis & Abbey Publishers

vi

development networks. His consulting and research interests are in Public Policy/ Finance, Economic Theory and Legislative Practice.

Dr Benedict Ndubisi Akanegbu is an Associate Professor of Economics at Nasarawa State University Keffi, Nigeria. He graduated with a First Class Honours degree in Economics from Alabama A & M University, USA and a PhD (Economics) from Howard, USA. He has over two decades of experience in teaching, research and community development. He has taught and researched in Howard University, USA, University of Lagos, the American University of Nigeria, Yola; and Nigerian Turkish Nile University, Abuja. He has also served in research and administrative capacities in several institutions including the World Bank's, Research and Evaluation Department, Washington D.C, and United States Department of Housing and Urban Development.

Robert Madu received his initial training in Journalism before reading for a Master's degree in Business with specialization in International Marketing, from the School of Business, Malardalen University, Sweden. Mr. Madu teaches courses on New Media, Media Economics and Marketing Communications at the Department of Mass Communication, Institute of Management and Technology, Enugu, Nigeria.

Dr Shedrack Moguluwa (PhD, Marketing), is a senior lecturer and former Head of Department of Marketing, University of Nigeria, Enugu campus. Dr Shedrack has published articles in many reputable journals.

Dr Justina Nnabuko is a Professor at the Department of Marketing, University of Nigeria, Enugu campus. She had been

List of Contributors   Jideofor Adibe (Ed.)
The Politics and Economics of Removing Subsidies on Petroleum Products in Nigeria
London & Abuja, Adonis & Abbey Publishers

vii

Head of Department and also Associate Dean of the Faculty of Business Administration in the same University. Professor Nnabuko has published books and articles in several reputable journals.

Dr. Abdelrasaq Nal holds a PhD in Economics from University of The Western Cape, Cape Town, South Africa. He currently serves as Head, Department of Economics & Development Studies, Federal University Dutsinma, Katsina State, Nigeria. Dr Nal is also an Associate of the South African Research Chair Initiative at the Pretoria- based Institute for Economics Research on Innovation., Dr. Nal has taught and researched widely in economics and related areas. He has published profusely and some of his works have been published by such reputable institutions as the Oxford University and United Nations University.

Evans S. OSABUOHIEN, (Ph.D, Economics) is a lecturer in the Dept. of Economics and Development Studies, Covenant University, Nigeria. He is currently a joint Alexander von Humboldt (AvH) Postdoctoral Researcher at GIGA German Institute for Global and Area Studies, Hamburg and German Development Institute, Bonn. He has won several awards and grants including the prestigious Alexander von Humboldt Fellowship for Postdoctoral Researchers, the Swedish Institute Guest Doctoral Fellowship, Council for Development of Social Science Research in Africa (CODESRIA)'s grant for Ph.D Thesis Writing; and Research Grant from Global Trade Alert (GTA) for The Centre for Economic and Policy Research (CEPR), UK. He has published over 30 scholarly articles in journals and edited books.

List of Contributors | Jideofor Adibe (Ed.)
The Politics and Economics of Removing Subsidies on Petroleum Products in Nigeria
London & Abuja, Adonis & Abbey Publishers

viii

Uchenna EFOBI is a doctoral candidate and a lecturer in the School of Business, Covenant University. He is also a qualified Chartered Accountant. His research focus is on Development, Institutions and International Economics. He has contributed chapters in edited books and published in reputable journals including the South African Journal of Economics, African Development Bank's Working Paper series, Emerald Book Series, IGI-Global Book Series and Petroleum-Gas University of Ploiesti Economic Sciences Series.

Ibukun BEECROFT is a doctoral candidate and an Assistant Lecturer and Researcher in the Department of Economics and Development Studies, Covenant University, Ota, Ogun State, Nigeria. She holds a B.Sc. in Economics from Covenant University and M.Sc. in Finance and Development from the School of Oriental and African Studies (SOAS), University of London, United Kingdom.

Before joining Covenant University, she worked with the Office for National Statistics, London, United Kingdom as well as with Zenith International Bank, Lagos, Nigeria. She has won various research and travel grants to conduct economic research including on 'Free Trade Versus Protectionism in Africa: The Unending Contradiction' (with Osabuohien, E. and Efobi, U.), funded by the Centre for Economic Policy Research (CEPR), United Kingdom, in collaboration with Global Trade Alert (GTA) and African Centre for Economic Transformation (ACET), Ghana. Her research paper (with two others) also won first place in the FLACSO-WTO Chairs Award in 2012. She is currently a reviewer for the *Journal of Environment, Development and Sustainability*. Her research interests include International Economics, Public Finance and Economic Development. She has

List of Contributors | Jideofor Adibe (Ed.)
The Politics and Economics of Removing Subsidies on Petroleum Products in Nigeria
London & Abuja, Adonis & Abbey Publishers

ix

participated in conferences and workshops within and outside Nigeria.

Yahaya Abdullahi Adadu holds a Master's degree in Political Science from the University of Jos, where he is also awaiting the final defence of his doctorate degree in the same field. He has been on the teaching staff of Nasarawa State University, Keffi since 2002, where he is Senior Lecturer and currently the Acting Head of the Department of Political Science. Adadu is widely published in reputable national and international journals. He has co-edited a book on Democracy and Development in Africa and has attended many conferences both locally and internationally.

Usman Abu Tom has been on the teaching staff of Nasarawa State University, Keffi since 2006. He holds a Master's degree in Political Science from Ahmadu Bello University, Zaria, and has completed his doctorate degree in Political Science at Nasarawa State University, Keffi. Usman, has published widely in the areas of Political Economy and Nigerian politics.

Hafsat Kigbu holds a Master's degree in Political Economy and Development Studies from the University of Jos and has been teaching Political Science at Nasarawa State University, Keffi since 2008. She has published in many national and international journals.

List of
Contributors

Jideofor Adibe (Ed.)
The Politics and Economics of Removing Subsidies on Petroleum Products in Nigeria
London & Abuja, Adonis & Abbey Publishers

X

INTRODUCTION

De-subsidizing PMS in Nigeria: Between the People, State and Oil Cabals

Jideofor Adibe

When in 2011 the federal government announced plans to remove subsidies from petroleum products, effective from January 2012, it was thought to be the latest in the seemingly endless stream of controversial policies that had been flowing from, and unwittingly increasingly defining the Jonathan regime that year. According to the supporters of de-subsidization, subsidizing Premium Motor Spirit (PMS), otherwise known as fuel in Nigeria, comes at enormous cost to the government, money, they argued, that could be best channelled to other sectors of the economy. For instance Moyo and Songwe (2012) argued that in 2011 alone subsidizing PMS cost the country an estimated $8 billion, with the price tag for 2012 expected to be substantially more, if the government had not substantially cut down on the subsidy from January that year. It is generally argued that the cost of subsidizing fuel grew exponentially partly due to its rising cost in the world market - which meant that the government had to spend even more to keep domestic prices low - and partly due to Nigeria's increasing population - which resulted in increased fuel consumption. Critics of subsidy argue that a combination of these factors made fuel subsidy unsustainable. For instance they argue that while the price of crude oil increased from 30.4 dollars per barrel in 2000 to 94.9 in 2010, over the same period, Nigeria's population increased from about 123 million to 158 million, meaning a significant increase in domestic demand and a concomitant increase in the total amount needed to maintain the subsidy regime. They contend that largely because of these factors, by 2011, fuel subsidy accounted for 30

Introduction   Jideofor Adibe , in
Jideofor Adibe (Ed.)
The Politics and Economics of Removing Subsidies on Petroleum Products in Nigeria
London & Abuja, Adonis & Abbey Publishers

per cent of the Nigerian government's total expenditure, which was about 4 per cent of GDP and 118 per cent of the capital budget (Moyo and Songwe, 2012).

Following from the above, proponents of de-subsidization contend that removal of subsidies will help the government to save some "N1.2tn, part of which can be deployed into providing safety nets for the poor segments of the society to ameliorate the effects of the subsidy removal", according to a letter to the National Assembly by the President conveying his administration's Medium Term Expenditure Framework (MTEF) and a N4.8tn budget for the 2012 fiscal year (cited in Adibe, 2012).

During the debates that preceded the government's decision to remove the fuel subsidies, there were varying estimates on how de-subsidization would affect the price of petroleum products. What was clear however that was following the announcement of the removal of fuel subsidies on January 1 2012, the official per litre price of petrol jumped from N65 to N141. On January 9 2012, barely a week after the de-subsidization announcement, 'shut-down Nigeria' strikes and protests spread across the country, truly shutting down economic and social life in the country. Given the depth of the fault lines in the country and their politicisation, the government, which did not expect that Nigerians could overcome their differences to sustain such a nationwide protest, was forced to negotiate with labour and civil society leaders. Both sides shifted ground – as had been the pattern in the politics of de-subsidization of PMS: while the government agreed to maintain a level of subsidy, labour and civil society leaders also shifted from their total opposition to any form of de-subsidization to accepting the need for a partial de-subsidization. The consequence was that the government adjusted the official per litre price from N141 to N97 – much higher than the N65 it was before the January 1 2012 announcement.

With 37.2 billion barrels of proven oil reserves, Nigeria has the second-largest reserves in Africa (after Libya) and is the continent's

Introduction	Jideofor Adibe , in
	Jideofor Adibe (Ed.)
	The Politics and Economics of Removing Subsidies on Petroleum Products in Nigeria
	London & Abuja, Adonis & Abbey Publishers

largest oil producer. Despite this, the country is the only member of the Organization of Petroleum Exporting Countries (OPEC) that imports refined fuel, with periodic scarcities. Prior to the latest round of reduction in the level of subsidy, the government argued that the per liter price of N65 was against a landing cost of N139. It also claimed that at that price it contributed a subsidy of N73 per liter or an annual total of N1, 200 billion (US$7.6 billion), or 2.6 per cent of the country's GDP (Akinbobola, 2012).

In the politics of de-subsidization under the Jonathan administration, Dr Ngozi Iweala, the Finance Minister, by virtue of her designation as the Co-ordinating Minister of the Economy Minister and her pedigree in the World Bank, was the face of the de-subsidization lobby. This provided a convenient cover for some high-ranking members of the administration and the PDP to find subtle ways of distancing themselves from the de-subsidization lobby without appearing to be opposing the President. For instance in a report on October 10, 2011 aptly entitled: 'Fuel subsidy: National Assembly, Okonjo-Iweala on collision course', Sweet Crude, which reviews Nigeria's energy industry, reported that "members of the National Assembly, especially Senators, were not satisfied with the MTEF and Okonjo-Iweala's alleged insistence that it would only take the removal of fuel subsidy for the framework to be successful" (Sweet Crude, 2011). The online report further noted that "even before the letter was submitted, the lawmakers were not happy with the minister's utterances which the legislators believed meant that her positions on economic policies should be final."

Ministers and other government functionaries who probably felt uncomfortable with the designation of Dr Iweala as Co-ordinating Minister for the Economy, a sort of Prime Minister, apparently found an opportunity to heap the blame for the unpopular policy on the Finance Minister. It was also obvious that Dr Iweala was not unaware of such a whispering campaign and had to defend herself. For instance the Punch of 19 January 2012 quoted the Finance Minister as

Introduction | Jideofor Adibe , in
Jideofor Adibe (Ed.)
The Politics and Economics of Removing Subsidies on Petroleum Products in Nigeria
London & Abuja, Adonis & Abbey Publishers

saying that she was not "behind the Federal Government's decision to remove fuel subsidy on January 1, 2012." Testifying at the hearing of the House of Representatives Ad Hoc Committee on Management of Fuel Subsidy, the minister was further quoted by the paper as saying: "The fuel subsidy policy removal is a government decision; not an Ngozi Okonjo-Iweala decision. People have been loading the blame of this policy on me, and I read them all over the Internet.... I repeat, the fuel subsidy removal is a government decision. I will not allow anyone to put the blame for something I didn't do on me" (Punch, 2012).

Despite Dr Iweala's denial, the truth is that her return as Finance Minister and *de facto* Prime Minister also coincided with a period of more strident push by the government for the removal of fuel subsidy. Dr Okonjo Iweala, as most people know, worked most of her adult life in the World Bank, where she rose to become one of its Managing Directors. The World Bank and its sister institution the International Monetary Fund are notorious for their brand of economics for the developing world, which almost always hinges on the removal of subsidies and the devaluation or flotation of the exchange rate regime (see Marchesi, and Thomas, 1999; Havnevik, 1988; Akonor, 2006; Williamson, 1983). In the 1980s and early 1990s, such policies, encapsulated in the structural adjustment programmes (SAPs), were forced on African countries by these two institutions and contributed in ruining African economies and reversing the developmental gains recorded since independence (Shah, 2013, Bernard, & Mengisteab 1993; Kapur 1998, Hanlon, 2003).

Remarkably many people who were against Dr Okonjo Iweala's appointment as Finance Minister hinged their opposition on fears that her economic prognoses could turn out to be a simple rehash, by reflex, of the discredited economic desiderata of the two Bretton Woods institutions.

It should be underlined that the proposal to remove fuel subsidies did not originate with the Jonathan administration. Debate over

Introduction	Jideofor Adibe , in
	Jideofor Adibe (Ed.)
	The Politics and Economics of Removing Subsidies on Petroleum Products in Nigeria
	London & Abuja, Adonis & Abbey Publishers

subsidies in general in fact date back to 1982 when President Shehu Shagari introduced the first set of austerity measures to address the then worsening economic situation faced by the Nigerian economy under his leadership (Abubakar, 2011). Shagari's austerity measures, influenced by the two Bretton Woods institutions, required the deregulation and removal of all subsidies from domestic prices of consumer products.

Between 1998 and 2009 alone, there were 11 adjustments in the pump price of fuel. While the late General Sani Abacha regime (1993 to 1998) pegged the pump price at N11 per litre throughout his five-year reign, the price was adjusted to N19 during the nine-month transitional administration of General Abdulsalam Abubakar. Former President Olusegun Obasanjo adjusted the price eight times (i.e. June 1, 2000; January 1, 2002, June 20, 2003; October 1, 2003; May 29, 2004; January 2005; August 2005 and May 25, 2007) during his eight-year reign between 1999 and 2007, from N19 per liter to N70 per liter, before the late President Musa Yar'Adua adjusted it downwards to N65 per liter. On August 15, 2011, based on its pricing template, the Petroleum Products Pricing Regulatory Agency (PPPRA) said that the landing cost of a liter of petrol in Nigeria was N129.2 while the margin for transporters and marketers was N15.49 per liter, which in its opinion meant that the export pump price ought to be N144.70 per liter instead of the N65 per liter that was charged.[1] Nigerians were ushered into the New Year on January 1 2012 with the announcement by the Goodluck Jonathan administration that fuel subsidies had been removed.

Threats of removing subsidies from PMS had been a tool of blackmail by various regimes in the country. The 'game' is often

---

[1] Not everyone accepted the argument that there was subsidy on petrol. For instance Professor Tam David-West, Petroleum Minister during the era of General Muhammadu Buhari (December 1983 – August 1985), and Mines, Power and Steel in the General Ibrahim Babangida government (August 27 1985- August 27 1993), contended that there was no oil subsidy in Nigeria, and that the government's contention that it is subsidizing fuel is a lie and a fraud (see Sahara Reporters, 2011).

Introduction | Jideofor Adibe , in
| Jideofor Adibe (Ed.)
| *The Politics and Economics of Removing Subsidies on Petroleum Products in Nigeria*
| London & Abuja, Adonis & Abbey Publishers

played this way: the regime in power proposes to remove fuel subsidies, all of it in one go, making a mountain of how much it costs the government and how the money saved from such removal would be used to transform the country into an el Dorado on earth. Organised labour and civil society come out fighting against the proposed removal. The government hardens its stance and organised labour and civil society also harden theirs. Some political, religious and traditional rulers are recruited to play the good cops by pleading for dialogue between the two contending sides. A dialogue subsequently ensues, which results in subsidy being removed by say 30 per cent, enabling labour and civil society leaders to claim some victory and the regime to achieve its original aim. This pattern became very predictable, especially under Obasanjo (1999-2007).

When the government complains about the high cost of subsidising petroleum products, often with affirmative nods from the two Bretton Woods institutions, it conveniently forgets that subsidies, in one form or the other, are facts of life in several countries across the world. The United States for instance currently pays around $20 billion per year to farmers in direct subsidies as "farm income stabilisation" measure (The Scottish Parliament, 2013; Akukwe, 2012). These subsidies had their origins in the US Farm bills, which date back to the economic turmoil of the Great Depression of the 1920s. Since then a tradition of government support to US farmers has been maintained. It is in fact estimated that for every dollar U.S. farmers earn, 62 cents come from some form of government subsidy. The estimated total subsidies to US farmers in 2009 from all levels of government were $180.8 billion (Nwachukwu, 2011).

Subsidy regime is also very strong in Europe. For instance the Common Agricultural Policy (CAP), a system of European Union subsidies, represented 48 per cent of the EU's budget of €49.8 billion in 2006. In 2010, the EU spent €57 billion on agricultural development, of which €39 billion was spent on direct subsidies (MercoPress, 2011; EUR-LEX, no date,). China has several export subsidies (Defever &

Introduction	Jideofor Adibe , in
	Jideofor Adibe (Ed.)
	The Politics and Economics of Removing Subsidies on Petroleum Products in Nigeria
	London & Abuja, Adonis & Abbey Publishers

Riaño, 2013). So the whole talk about how much subsidies cost the government is like moaning about how much you spend to keep your children healthy and happy. This is not to suggest that government must subsidize everything but rather a need for an appreciation that subsidy, direct or indirect, is a fact of life in many countries – including in many of the countries that have been the most vocal critics of Nigeria's subsidy regimes. In oil producing country like Nigeria, whose economy more or less runs on a generator, ensuring that the prices of fuel are low and affordable to ordinary citizens will have ramifications in other sectors of the economy such as on the cost of transportation and food prices.

The table below shows pre-tax subsidies for petroleum products for some oil producing countries. The table shows that subsidizers include wealthy countries like Saudi Arabia, poor ones like Yemen, and many in between. Oil exporters are especially prominent on the list, but there are importers there as well. However some major oil exporters such as Russia and the United Arab Emirates engage in little or no subsidization.

Table 1: Petroleum product Subsidies in selected countries

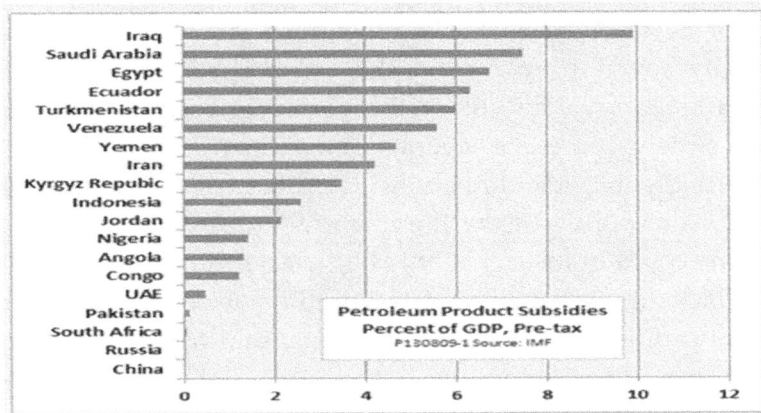

Source: Ed Dolan, 2013

Introduction | Jideofor Adibe , in
Jideofor Adibe (Ed.)
The Politics and Economics of Removing Subsidies on Petroleum Products in Nigeria
London & Abuja, Adonis & Abbey Publishers

Table 2: Per liter cost of fuel and diesel in selected oil producing countries

NATION		CURRENCY (US Dollar)
Algeria;	(Petrol) $0.41/litre	(Diesel)$0.20/litre
Bolivia;	(Petrol)$0.54/litre	(Diesel)$0.54/litre
Egypt;	(Petrol) $0.31/litre	(Diesel)$0.27/litre
Guatemala;	(Petrol)$0.64/litre	(Diesel)$0.54/litre
Indonesia;	(Petrol)$0.59/litre	(Diesel)$0.59/litre
Iraq;	(Petrol) $0.38/litre	(Diesel)$0.34/litre
Libya;	(Petrol)$0.17/litre	(Diesel)$0.15/litre
Malaysia;	(Petrol)$0.61/litre	(Diesel)$0.56/litre
Nigeria;	(Petrol) $0.41/litre	(Diesel) $1.00/litre
Venezuela;	(Petrol)$0.04/litre	(Diesel)$0.03/litre

Source: Adekunle, Oseni Semiu (2013)

Again by emphasising on the quantum of money to be saved from the de-subsidization of PMS, the government wrongly implied that lack of money is the only cause of the current problems facing the country. If lack of money is really the only or even the key problem, one might be tempted to ask what impact the money the country purportedly saved from exiting the sovereign debts owed to the London and Paris Clubs under the Obasanjo regime made on the material circumstances of average Nigerians? Has any ordinary Nigerian really felt the impact of the billions of Naira allegedly recovered from politicians by the financial crimes busters, EFCC and ICPC? One could in fact argue that Nigeria's fundamental problem is less the lack of money than it is of the vision thing and being repeatedly saddled with leaders who are either not prepared for the job or are incapable of thinking outside the established dogmas and wrong policies that have failed in the past. As part of the social contract in every country, the citizens must have something to show for giving up their right to self-help and giving loyalty to a central

Introduction | Jideofor Adibe , in
Jideofor Adibe (Ed.)
The Politics and Economics of Removing Subsidies on Petroleum Products in Nigeria
London & Abuja, Adonis & Abbey Publishers

authority. Without anything substantial from the government to justify itself in the eyes of the people, many citizens will willy-nilly begin to withdraw from the state and transfer their loyalty to any primordial identity where they feel they can best negotiate the meaning of their existence.

It may also be necessary to interrogate the claims by the de-subsidization lobby that petrol is too cheap in the country (see table 2 above). The truth is that at N65 per litre, it is actually not, especially given the country's per capita income. In the UAE and Saudi Arabia for instance, where per capita income is several times higher than what you have in Nigeria, a litre of petrol sells for around 1AED (N45) and SRO 0.45 (N18) respectively (Nwachukwu, 2011). In these countries, the citizens additionally enjoy their God-given natural wealth in several other ways. As one of the largest oil producing and exporting countries in the world, one may be tempted to ask in which way does citizen Okeke, Musa or Banjo feel the impact of the wealth that crude oil brings to the country.

There is no doubt that the partially botched attempt to completely remove subsidies on PMS on January 1 2012 and the protests that followed added to the legitimacy crisis that appears to be engulfing the Jonathan administration from all corners as one controversial policy feeds into the other amid pervasive insecurity and rising poverty. An increasing loss of faith in the ability of the regime to guarantee the security of life and property conflates with an unprecedented crisis in the country's nation building project, unacceptably high level of unemployment and incessant strikes by organized labour, including members of the Academic Staff Union of Universities, to create an escalating crisis of legitimacy for the regime.

Debate over subsidization of PMS also revealed how a cabal among oil marketers had been reaping off the country. For instance probes by both the House of Representatives and federal government- appointed committees revealed massive abuse of the

Introduction	Jideofor Adibe , in
	Jideofor Adibe (Ed.)
	The Politics and Economics of Removing Subsidies on Petroleum Products in Nigeria
	London & Abuja, Adonis & Abbey Publishers

subsidy regime by oil marketers. On 25 July 2012 for instance, This Day newspaper reported:

> EFCC named the suspects in the massive fuel subsidy frauds to include seven oil companies and 12 individuals who have been indicted during investigations into the management of the fuel subsidy scheme and would be arraigned before judges of the Lagos State judiciary. The suspects, comprising seven oil companies and 12 individuals, are: Nasaman Oil Services; Eterna Oil and Gas Plc; Ontario Oil and Gas Plc; Nadabo Energy Limited; Pacific Silver Line Limited, Axenergy Limited and Fago Petroleum and Gas Limited.
>
> The 12 individuals involved in the scam are: Mamman Nasir Ali, Christian Taylor, Mahmud Tukur, Ochonogor Alex, Walter Wagbatsoma, Adaoha Ugo-Ngadi, Fakuade Babafemi Ebenezer, Ezekiel Olaleye Ejidele, Abubakar Ali Peters, Jude Agube Abalaka, Abdullahi Alao and Oluwaseun Ogunbanbo
>
> Mamman Nasir Ali, who runs Nasaman Oil Services, is the son of former PDP National Chairman, Col. Ahmadu Ali (rtd), who was also the chairman of the Petroleum Products Pricing and Regulating Agency (PPPRA) board.
>
> Mahmud Tukur, who runs Eterna Oil and Gas Plc, is the son of the current PDP National Chairman, Alhaji Bamanga Tukur, while Abdullahi Alao of Axenergy Limited is the son of popular Ibadan-based tycoon, Alhaji Abdulazeez Arisekola-Alao.
>
> Ugo-Ngadi, on her part, is the Managing Director of Ontario Oil and Gas Plc, where its founder, Wagbatsoma, is the executive vice-chairman (ThisDay, 2012).

The revelation of massive fraud in the fuel subsidy regime by a cabal of oil marketers hardened the position of those opposed to de-subsidization as they argued that the problem was not with fuel subsidy but with the government not having the necessary political will to fight the fraudsters. It coincidentally also strengthened the resolve of those opposed to subsidy as they in turn argued that the marketers are so powerful and so well connected that the only way to fight the fraud in the fuel subsidy regime is to completely remove the subsidies.

Introduction | Jideofor Adibe , in
Jideofor Adibe (Ed.)
The Politics and Economics of Removing Subsidies on Petroleum Products in Nigeria
London & Abuja, Adonis & Abbey Publishers

The contributions in this volume address the various aspects of the de-subsidization arguments from different theoretical, disciplinary and ideological perspectives. Emmanuel Ojameruaye in 'The Evolution and Political Underpinnings of the Petroleum Products Subsidy Debate in Nigeria' argues that "the current approach to the pricing of petroleum products in Nigeria is inefficient, ineffective and unsustainable" but added that the government is in a quagmire about doing the' right thing' because of fear of violent protests and opposition by labour unions and civil society groups

Ogujiuba Kanayo, in 'Petroleum Subsidy Debate in Nigeria: Issues and Policy Options' disputes the contention that deregulation and subsidy removal would encourage the private sector to invest in refineries in Nigeria. He argues that "even after the deregulation, private companies can still refuse to build refineries within the country but rather continue to import refined products" because of the non-sustainability of policies by governments. He further argues that "no private company will want to invest millions of dollars to build refineries, only for the pipelines to be blown-up, employees taken hostage or even worst, killed". For him therefore, addressing the questions of sustainability of policies, insecurity, political instability and dilapidated infrastructure should be precedent to deregulation of the downstream sector. He advocates the sequencing of de-regulation and a dialogue between the federal government, organized labour and other stakeholders before any further deregulation of the petroleum downstream sector.

In 'A case for Fuel De-subsidization in Nigeria' Benedict Ndubisi Akanegbu makes an unapologetic case for the removal of subsidy on PMS. He examines the economic effects of the deregulation of the downstream oil sector and puts a proviso that subsidies can be justified only if it can increase overall social welfare – something he argues does not happen under the current subsidy regime. His position is that "experience has shown that net effects of subsidies are

Introduction	Jideofor Adibe , in
	Jideofor Adibe (Ed.)
	The Politics and Economics of Removing Subsidies on Petroleum Products in Nigeria
	London & Abuja, Adonis & Abbey Publishers

21

negative. And this means that the overall social welfare would be higher without subsidies".

Robert Madu and Shedrach Moguluwa took an opposite theoretical stance from Benedict Akanegbu. In 'Subsidy as an imperative for sustainability in a depressed economy: a case study of Nigeria', the duo made a case for the subsidization of PMS, contending that in global and comparative terms, countries and regions have used subsidies for their citizens and members over the years such as the current bailout of Greece by the European Union. They further argued that subsidy removal will leave the Nigerian economy in a more depressed state than it met it considering the complexities and peculiarities surrounding the Nigerian polity.

Emmanuel Ojameruaye in 'Petroleum Products subsidies in Nigeria: Economic and Global Perspectives' notes the difficulty of estimating the actual level of PMS subsidy in the country arguing that revelations of the massive fraud in the management of fuel subsidies by oil marketers showed that corruption accounted for a "substantial part of the increase in the amount of fuel subsidy. The House of Representative Ad Hoc Committee on the Fuel Subsidy Report revealed that the astronomical increase in subsidy payment was due to corrupt practices." He also contends that the fact that subsidies are Pareto inefficient does not necessarily justify their removal or reduction. For him:

> As long as a government imposes consumption tax (such as value added tax or sales tax on some commodities), it can be argued that the government should also subsidize certain commodities, especially those that are "critical" to the economy and those that benefit the poor more than the rich. Thus, while many economists will argue that the elimination of fuel subsidy is good for the economy and is a "win-win" or "no lose" situation at least from a theoretical point of view, others have embraced the political and social realities that either justify fuel subsidies or make their elimination impracticable, at least in the short to medium terms.

Introduction	Jideofor Adibe , in
	Jideofor Adibe (Ed.)
	The Politics and Economics of Removing Subsidies on Petroleum Products in Nigeria
	London & Abuja, Adonis & Abbey Publishers

Abdelrasaq Nal in 'Inclusive Growth and the Contradiction of Petroleum Product De-subsidisation Strategy' interrogates the 'inclusive growth framework' which begins with the assumption that at any point in time, and in any economy, various forms of distortions exist. He argues that dealing with these distortions in a welfare maximising manner requires a rethink of the government's current approach to policy reform in a number of areas. For instance he challenges the government argument that de-subsidization would lead to greater investment in the downstream sector by the private sector. As he puts it:

> Perhaps the best way to dismiss the government's claim is to ask: why is it that despite the absence of subsidies in other sectors like manufacturing, private investments have remained desperately low? If the truth must be told, private investment issue in Nigerian refineries extends far beyond subsidy. It touches on wider fundamental problems of corruption and generally poor investment climate in the country.

Efobi Uchenna, Osabuohien Evans & Beecroft Ibukun in 'The Macroeconomic Consequences of the Black Sunday in Nigeria' examine the effects of the change in pump price of fuel on exchange rate, inflation and money supply. Their reason for focusing on the three macroeconomic indicators is because "they can react very quickly to any shock such as de-subsidization." They were able to establish a reaction of these variables to the reduction in the fuel subsidy regime and from this concluded that

> ... fuel price is a very sensitive variable which can alter the macroeconomic equilibrium of an economy. Therefore the government should have an understanding of this when deliberating on future policies. This will help in reducing socio-economic tensions of policies as the populace will need some time to adjust to policy changes. Despite the government's 'good' intentions for policy change such as the substantial cut in the fuel subsidy, when the process lacks proper timing, the macroeconomic consequences can be gruesome.

Introduction | Jideofor Adibe , in
Jideofor Adibe (Ed.)
The Politics and Economics of Removing Subsidies on Petroleum Products in Nigeria
London & Abuja, Adonis & Abbey Publishers

Adadu Yahaya Abdullahi, Usman Abu Tom and Hafsat Kigbu, in 'Petroleum products pricing and the manipulation of oil subsidy in Nigeria' examine the issues involved in the pricing of petroleum products and removal of subsidies from 1999-2012. Their aim was to capture the phenomenon of incessant fuel crises in Nigeria. They blame these crises on vastly expanded energy consumption, irregular maintenance of the refineries which often result in their total breakdown, large scale smuggling of petroleum products to neighbouring countries enhanced by corrupt government officials and the role played by saboteurs behind the constant vandalization of pipelines and incessant strikes by oil workers, mainly those under the aegis of National Union of Petroleum and Natural Gas Workers (NUPENG) and the Petroleum and Natural Gas Senior Staff Association of Nigeria (PENGASSAN). They also argue that the government is not sincere to the citizens, noting for instance that while OPEC gave a concession to the Federal Government to exceed its quota by 445,000 barrel a day to provide for the domestic needs of Nigerians, government functionaries rather found that it would be more lucrative to them to cripple "the four refineries through neglect so as to continue selling this allocation internationally and making huge profits while using part of the proceeds to import refined products in a very non-transparent way, where the importers bring in cheap and loaded petrol with no special additives."

In the concluding part of the book, Jideofor Adibe examines the impact of the substantial reduction in the subsidy level of PMS against some of the arguments of the pro-de-subsidization lobby.

There are deliberate overlaps in some of the themes explored in this volume. The aim is to allow the examination of the same theme from different theoretical, ideological and even disciplinary perspectives. This will hopefully offer the reader a more holistic way of looking at these issues.

Introduction | Jideofor Adibe , in
Jideofor Adibe (Ed.)
The Politics and Economics of Removing Subsidies on Petroleum Products in Nigeria
London & Abuja, Adonis & Abbey Publishers

References

Abubakar N.K (2011) 'The Return of Fuel Subsidy Removal Debate', *Business Day*, 15 July

Adekunle, Oseni Semiu (2013) 'The impact of subsidy removal on Nigerian economy: series two', The National Economic Transformation' blog, 18 March, http://nationaleconomictransfor mation.blogspot.com/2013/03/the-impact-of-subsidy-removal-on_340.html (Accessed 30 July 2013).

Adibe, J (2011): 'The Politics of de-subsidisation', *Daily Trust*, November 10, (back page).

Akinbobola, Yemisi (2012): 'Bid to end subsidy stirs protest in Nigeria: Unrest highlights problems of mismanagement and corruption' *African Renewal* online, April http://www.un.org/africa renewal/magazine/april-2012/bid-end-subsidy-stirs-protest-nigeria (Accessed 20 September 2013)

Akonor, Kwame (2006): *Africa and IMF Conditionality: The Unevenness of Compliance, 1983-2000,* African Studies, (Oxford, Routledge)

Akukwe, Obinna (2012): 'Nigeria: Food Security and Problem of Middleman', 9 March, *AllAfrica.com,* http://allafrica.com/stories/20 1203120984.html (Accessed 20 July 2013)

Anise, L. (1980), 'De-subsidization: An Alternative Approach to Government Cost Containment and Income redistribution Policy in Nigeria', *African Studies Review*, vol. 23, No 2.

Defever, Fabrice & Alejandro Riaño, (2013), 'China's pure exporter subsidies: Protectionism by exporting' January 4, *Vox* http://www .voxeu.org/article/china-s-pure-exporter-subsidies-protectionism-exporting (Accessed 17 May 2013)

Dolan, Ed (2013): 'Why Fuel Subsidies are Bad for Everyone', 13 August, *OilPrice.com,* http://oilprice.com/Energy/Gas-Prices/Why-Fuel-Subsidies-are-Bad-for-Everyone.html (Accessed 10 October 2013)

EUR-LEX (no date), '2010 General Budget: Title 05 – Agriculture and Rural Development', http://eur-lex.europa.eu/budget/data/D2010_

Introduction	Jideofor Adibe , in
	Jideofor Adibe (Ed.)
	The Politics and Economics of Removing Subsidies on Petroleum Products in Nigeria
	London & Abuja, Adonis & Abbey Publishers

VOL4/EN/nmc-titleN123A5/index.html   (Accessed,   November 2012)

Hanlon, Joseph (2003): *Peace without Profit: How the IMF Blocks Rebuilding in Mozambique (African Issues)*, (London, James Currey)

Ikubolajeh, Bernard Logan aand Kidane Mengisteab (1993): "IMF-World Bank Adjustment and Structural Transformation on Sub-Saharan Africa". *Economic Geography.* Vol 69, No 1, African Development.

Kapur, Davesh. (1998). The IMF: A Cure of Curse? *Foreign Policy.* No. 111, 114-129.

Kjell J. Havnevik, 1988, ed.: *The International Monetary Fund and the World Bank in Africa: Conditionality, Impact and Alternatives - Seminar Proceedings* (Uppsala, Nordia Africa Institute).

Marchesi, S and Thomas, J.P. (1999) 'IMF Conditionality as a Screening Device', *Economic Journal* 109, 111-125.

Meltzer, A.H. (2006): 'Reviving the Bank and the Fund' *Review of International Organizations, 1, 49-59.*

MercoPress (2011): 'EU farm ministers support strong CAP with "proportional financial resources"', March 21, http://en.mercopres s.com/2011/03/21/eu-farm-ministers-support-strong-cap-with-proportional-financial-resources(Accessed, August 20 2013)

Moyo, Nelipher and Vera Songwe (2012): 'Removal of Fuel Subsidies in Nigeria: An Economic Necessity and a Political Dilemma' Brookings, http://www.brookings.edu/research/opinion s/2012/01/10-fuel-subsidies-nigeria-songwe(Accessed   20   April 2013).

Nwachukwu, Kanayo (2011): 'Removing petrol subsidies in Nigeria is senseless', http://chiedufelix.blogspot.com/2011/11/removing-petrol-subsidies-in-nigeria-is.html, November 22 (Accessed   24 September 2013)

Punch, (2012): 'I'm not behind removal of fuel subsidy – Finance minister', 19 January, http://www.punchng.com/news/i-dont-

Introduction | Jideofor Adibe , in
Jideofor Adibe (Ed.)
The Politics and Economics of Removing Subsidies on Petroleum Products in Nigeria
London & Abuja, Adonis & Abbey Publishers

know-if-my-ministry-is-on-pppra-board-finance-minister/ (Accessed, 23 September 2013)

Sahara Reporters (2011): 'Oil Subsidy is Fiction - Okonjo Iweala is here to seduce people to accept the callous "oil subsidy" removal-Tam David West-TheNEWS', October 28, http://saharareporters.c om/interview/oil-subsidy-fiction-okonjo-iweala-here-seduce-people-accept-callous-oil-subsidy-removal-ta (Accessed, 1 October 2013).

Shah, Anup (2013): "Structural Adjustment—a Major Cause of Poverty." *Global Issues*, March 24, <http://www.globalissues.org/ar ticle/3/structural-adjustment-a-major-cause-of-poverty (Accessed 20 September 2013).

Sweet Crude (2012): 'Okonjo-Iweala, lawmakers on collision course over fuel subsidy, *Sweet Crude*, October 10, http://sweetcruderepor ts.com/2011/10/10/okonjo-iweala-lawmakers-on-collision-course-over-fuel-subsidy/ (Accessed 20 September 2013)

The Scottish Parliament (2013): 'Official Report Debate Contributions', 13 March, http://www.scottish.parliament.uk/parli amentarybusiness/28862.aspx?r=7813&i=71251&c=1432958 (Accessed 2 October 2013).

This Day (2012): 'Subsidy Fraud: EFCC to Prosecute 23 Oil Marketers',http://www.thisdaylive.com/articles/subsidy-fraud-efcc-to-prosecute-23-oil-marketers/120779/, July 25 2012 (accessed 11October 2013)

Vogel, Stephen (2012): "Farm Income and Costs: Farms Receiving Government Payments", Ers.usda.gov. (Accessed, 10 November 2012).

Williamson, John (1983): *IMF Conditionality* (Massachusetts, MIT Press)

Introduction | Jideofor Adibe , in
Jideofor Adibe (Ed.)
The Politics and Economics of Removing Subsidies on Petroleum Products in Nigeria
London & Abuja, Adonis & Abbey Publishers

CHAPTER ONE

The Evolution and Political Underpinnings of the Petroleum Products Subsidy Debate in Nigeria

Emmanuel Ojameruaye

This chapter examines the evolution of petroleum products subsidy in Nigeria and the politics surrounding the debate to eliminate the subsidy.

The Evolution of Petroleum Products Subsidy in Nigeria

The pricing of petroleum products in Nigeria has been a bane to the Nigerian economy because petroleum product prices are fixed and subsidized by the Federal Government (FG). In addition, the petroleum products market in Nigeria is characterized by inefficiencies, frequent disequilibria and massive corruption. The subsidy on petroleum products in Nigeria increased phenomenally from N261.1 billion (US$2.03 billion) in 2006 to N633.2 billion (US$5.37 billion) in 2008 and reached N2, 188 billion ($14.12 billion) in 2012 – representing about 6% of the country's GDP, 20% of the gross Federation Account revenue, 46% of the aggregate expenditure of the FG and more than double the Capital Expenditure of the FG (Adenikinju, 2009; CBN, 2012). Subsidy payment to importers of petroleum products and the Nigerian National Petroleum Corporation (NNPC) that operates the local refineries is in fact crowding out investment in other vital areas such as infrastructure, education and health. Figure 1 shows the phenomenal increase in subsidies in recent years.

The increase in subsidy is due to a combination of factors including increase in the consumption of petroleum products, inadequate domestic production of petroleum products by local

Chapter one | Emmanuel Ojameruaye, in
Jideofor Adibe (Ed.)
The Politics and Economics of Removing Subsidies on Petroleum Products in Nigeria
London & Abuja, Adonis & Abbey Publishers

29

(federal-government owned) refineries, increase in the quantity of imported petroleum products to cope with increasing demand, increase in the price of imported petroleum products due to the increase in crude oil prices in the world market, non-adjustment of the administered prices of petroleum products to match supply costs and massive corruption associated with the subsidy regime. In fact, the increases in the retail price of petroleum products over time have lagged far behind the increases in the supply cost of petroleum products. Since petroleum products are internationally traded goods and crude oil accounts for a significant proportion of the supply cost of petroleum products, we can use the price of crude oil as a proxy for the supply cost of petroleum products.

1: **Evolution of Petroleum Products Subsidy in Nigeria, 2006-2012**

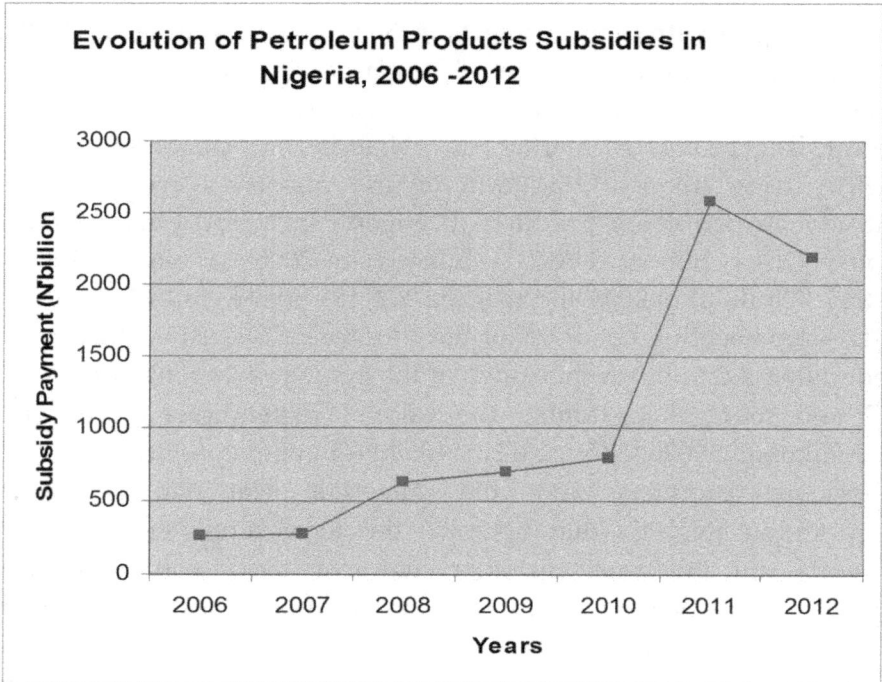

Source: Data for 2006-2008 are from Adenikinju (2009) while data for 2009 -2010 are estimates and data 2011 and 2012 are from FMF website

Chapter one | Emmanuel Ojameruaye, in
Jideofor Adibe (Ed.)
The Politics and Economics of Removing Subsidies on Petroleum Products in Nigeria
London & Abuja, Adonis & Abbey Publishers

Figure 2 shows the evolution of the price of premium motor spirit (PMS) otherwise known as petrol or gasoline which account for about 78% percent of the petroleum products consumed in Nigeria. The other key petroleum products are kerosene (dual purpose kerosene or DPK) which account for about 10% and diesel (automotive gas oil or AGO) which accounts for about 12%. The chart shows that gasoline price has increased significantly over the past two decade, although the rate of increase is less than the rate of increase in the world (spot) market prices of gasoline and crude oil as shown in figure 3. The same is true for kerosene. However, as figure 3 shows, the same is not true for diesel because diesel price was deregulated in 2009 (i.e. subsidy on diesel was removed) thus ensuring that diesel prices reflect supply cost. In fact, figure 2 indicates that the increase in the retail price of diesel in Nigeria in recent years was slightly more than that of diesel and crude oil in the world market.

Most economists agree that the current approach to the pricing of petroleum products in Nigeria is inefficient, ineffective and unsustainable. However, several attempts to eliminate the subsidy and deregulate the market have failed as a result of violent protests and opposition by labor unions and civil society groups. It is an understatement to state that the government is currently at a crossroad on the issue of subsidy. Over the past 30 years, successive governments in Nigeria have grappled with the problem of petroleum products subsidy. Like most other public policy issues, there seem to be no generally acceptable solution because of the interplay of politics and economics. The subsidy issue has become a major political and economic headache for the Nigerian government because the current level of subsidy is unsustainable and the subsidy regime is riddled with corruption.

Chapter one | Emmanuel Ojameruaye, in
Jideofor Adibe (Ed.)
The Politics and Economics of Removing Subsidies on Petroleum Products in Nigeria
London & Abuja, Adonis & Abbey Publishers

Figure 2: Evolution of the price of Petrol in Nigeria

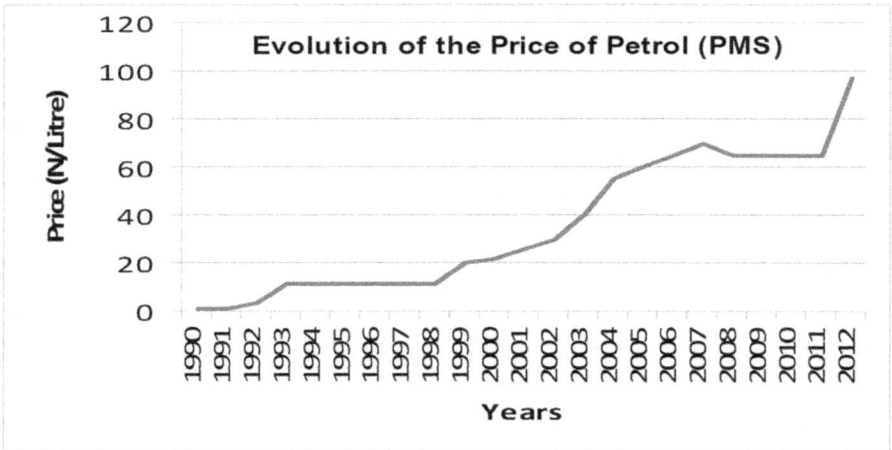

Figure 3: Indices of the Retail Prices of Gasoline and Diesel in Nigeria vis-à-vis Indices Spot Market (Rotterdam) Prices of Gasoline, Diesel and Nigeria Crude Oil.

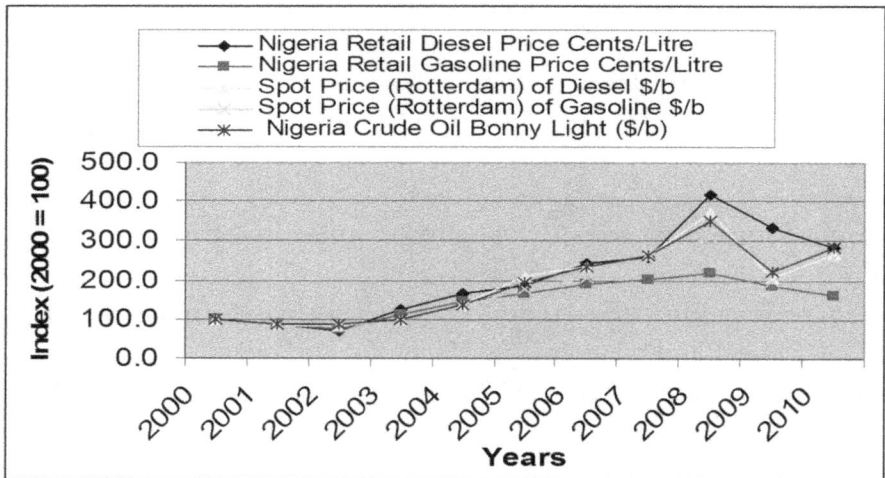

Source: Charted with price data from GIZ (2012) and OPEC Annual Statistical Report (2011). The Nigerian Retail Prices of gasoline and diesel are in US cents per litre, while the Spot Market prices are in US$ per barrel. All the prices were converted to index numbers (with 2000 = 100).

Chapter one | Emmanuel Ojameruaye, in
Jideofor Adibe (Ed.)
The Politics and Economics of Removing Subsidies on Petroleum Products in Nigeria
London & Abuja, Adonis & Abbey Publishers

Recent Attempts to Eliminate Petroleum Products Subsidy in Nigeria

In order to ensure that petroleum product prices reflect supply cost and the forces of demand and supply, the government embarked on the deregulation of the downstream oil industry in August 2000. A special committee on the Review of Petroleum Products Supply and Distribution was set up which submitted its report in October 2000. The government issued its White Paper on the report in January 2001 and the President forwarded the Bill for an Act to Establish Petroleum Products Pricing Regulatory Committee (PPPRC) to the National Assembly in March 2001. In January, 2002, the government commenced the liberalization of the downstream sector of the oil industry by setting the prices for petrol (PMS), diesel (AGO) and kerosene (HHK) at N26, N26 and N24 per litre, respectively. An import duty N1.50 per litre on imported petroleum products was introduced while selling price of crude to local refineries was increased from $9.50 to $18.0 per barrel. The Senate passed the Petroleum Products Pricing Regulatory Agency (Establishment) Bill on February 5, 2003 and the House of Representatives did the same on May 22, 2003. President Obasanjo signed the Petroleum Products Pricing Regulatory Agency (PPPRA) Bill into law on May 27, 2003 and inaugurated the Governing Board of the PPPRA on June 19, 2003.

The mission of the PPPRA is to "reposition Nigeria's downstream sub-sector for improved efficiency and transparency", and its vision is to "to attain a strong, vibrant downstream sub-sector of the petroleum industry, where refining, supply, and distribution of petroleum products are self-financing and sustaining" (www.pppra-nigeria.org).

The functions of the PPPRA are to:

a) Determine the pricing policy of petroleum products;
b) Regulate the supply and distribution of petroleum products;

Chapter one | Emmanuel Ojameruaye, in
Jideofor Adibe (Ed.)
The Politics and Economics of Removing Subsidies on Petroleum Products in Nigeria
London & Abuja, Adonis & Abbey Publishers

c) Create an information databank through liaison with all relevant agencies to facilitate the making of informed and realistic decisions on pricing policies;

d) Moderate volatility in petroleum products prices, while ensuring reasonable returns to operators;

e) Maintain constant surveillance over all key indices relevant to pricing policy and periodically approve benchmark prices for all petroleum products;

f) Identify macro-economic factors with relationship to prices of petroleum products and advice the Federal Government on appropriate strategies for dealing with them;

g) Prevent collusion and restrictive trade practices harmful in the sector; and

h) Create firm linkages with key segment of the Nigerian society, and ensure that its decision enjoy the widest possible understanding and support.

The PPPRA announced new pump prices of petroleum products on June 20, 2003 which was greeted by a nation-wide strike declared by the Nigerian Labour Congress (NLC) and its affiliates. The nation-wide strike ended with the adjustment of the prices to N34 per litre for petrol, N32 each for diesel and kerosene but retained subsidies on these products. The government eliminated subsidy on diesel and low pour fuel oil (LPFO) in August 2009 but retained subsidy on petrol and kerosene. On January 1, 2012, the federal government attempted to eliminate the subsidy on petrol and kerosene by increasing the price of petrol from N65 to N141 per litre. The action was greeted by violent protests organized by the NLC and the Trade Union Congress (TUC) and the government was forced to re-introduce the subsidy by fixing the price of petrol at N97 per litre while the price of kerosene was retained at N50 per litre.

Following the violent protests that greeted the announcement of the removal of the subsidy on January 1, 2012, the House of Representatives held an emergency meeting on January 8, 2012

Chapter one | Emmanuel Ojameruaye, in
Jideofor Adibe (Ed.)
The Politics and Economics of Removing Subsidies on Petroleum Products in Nigeria
London & Abuja, Adonis & Abbey Publishers

during which it set up an Ad-Hoc Committee, led by Alhaji Farouk Lawan, "to verify and determine the actual subsidy requirements and monitor the implementation of the subsidy regime in Nigeria" between 2009 and 2011. The Committee went to work immediately and submitted its report in April, 2013. The report revealed that the subsidy regime was characterized by massive corruption and a huge scam in which importers of petroleum products were paid hundreds of millions of naira in subsidies by the government for products not delivered. The Committee recommended that the NNPC, petroleum products importers and marketers should refund about N1.1 trillion to the Federation Account. It also called for further investigation of several companies for corrupt practices associated with the subsidy regime (Ad Hoc Committee, 2012). There were several technical issues with the report (Ojameruaye, 2012a). It was also revealed that the Chairman and Secretary of the Committee solicited for and accepted bribes from one of the companies investigated in order to remove the company from the list of companies that the committee had implicated in the scandal. In February 2013, the federal government charged the Chairman of the Committee Farouk Lawan and the Secretary to the committee Boniface Emenalo to court for allegedly obtaining $620,000 (N97 million) bribe from Femi Otedola, the Chairman of Zenon Oil and Gas Ltd. The case is still in court as at the time of finalizing this chapter (October 2013).

About ten years after it was established, the PPPRA is yet to achieve its mission and objectives. The major achievement of the PPPRA is the development of a petroleum products pricing template (PPPT) which it uses to determine the supply price of imported petroleum products and the level of subsidy derived as the difference between the supply price and the pump price fixed by the government. It is therefore safe to say that the deregulation policy which started over twelve years ago has not achieved its objectives. The prices of petroleum products are still administered and Nigeria is increasingly relying on imported petroleum products while the existing refineries are producing at less than 30% of their installed

Chapter one | Emmanuel Ojameruaye, in
Jideofor Adibe (Ed.)
The Politics and Economics of Removing Subsidies on Petroleum Products in Nigeria
London & Abuja, Adonis & Abbey Publishers

capacity. In fact, the cost of importing petroleum products has increased so rapidly in recent years that it is threatening the country's balance of payments and capital expenditures. This is what has given rise to the recurring debate and controversy over the removal of fuel subsidy. For instance, in the letter conveying the 2012-2015 Medium Term Expenditure Frameworks (MTEF) and 2012 Fiscal Strategy Paper to the National Assembly on September 22, 2011, President Jonathan stated that:

> A major component of the policy of fiscal consolidation is government's intent to phase out the fuel subsidy beginning from the 2012 fiscal year. This will free up about N1.2 trillion in savings, part of which can be deployed into providing safety nets for poor segments of the society to ameliorate the effects of the subsidy removal (*Daily Sun*, October 5, 2011).

That proposal quickly ignited a fierce debate. On the one hand, the representatives of government insisted on the phasing out or removing fully the subsidy on petroleum products. On the other hand many non-governmental individuals (NGI) and civil society organizations (CSO) and labour unions vehemently opposed the idea. Again, while presenting its report to President Jonathan on November 2, 2012, the Dr. Kalu-led National Refineries Special Task Force noted that

> The uniform and regulated pricing policy for petroleum products ... is one of the most widely adduced reasons...for lack of investment in new refineries in Nigeria... also believed to be substantially responsible for waste, distortions and corrupt practices in the industry". The task force therefore recommended that "It will therefore be necessary to fully deregulate prices in the Downstream Sector prior to the completion of the privatization process (Kalu, 2012).

Shortly after that, while receiving the report of the graduating participants of the Senior Executive Course of the National Institute of Policy and Strategic Studies on November 15, 2012, President Jonathan stated categorically that his government was bent on getting rid of the subsidy. Echoing the report of the Task Force, the President

Chapter one | Emmanuel Ojameruaye, in
Jideofor Adibe (Ed.)
The Politics and Economics of Removing Subsidies on Petroleum Products in Nigeria
London & Abuja, Adonis & Abbey Publishers

stated: "Why is it that people are not building refineries in Nigeria despite the fact that it is a big business? It is because of the policy of subsidy, and that is why we want to get out of it". However, in a swift reaction the next day, on November 16, the President of the Nigerian Labour Congress (NLC), Abdulwahid Omar, fired back at President Jonathan, saying

> Such a statement coming from... the President is highly disturbing and we want to believe that he will not toy with the tempers of Nigerians...coming at this time of the year, when many see the current fuel scarcity being experienced in most parts of the country as being artificially created (*Vanguard*, November 17, 2012)

This was a signal that the NLC would resist any move to remove or reduce the subsidy as it did in January 2012. Thus, the federal government was careful not to tamper with the subsidy in early 2013 as it did in 2012, and petroleum products have remained at their 2012 level as of April 2013, but most observers believe that it is only a matter of time before the prices will be increased by the government.

A major problem for the federal government is that it has not done a good job in presenting or "selling" the case for the removal of the subsidy to the National Assembly and the general public. Furthermore, most Nigerians including labor union leaders do not have a good knowledge and reliable data on the economics of petroleum products subsidy to make an informed decision on the issue.

The Politics of the Petroleum Products Subsidy Debate

In most cases around the world, subsidies are rooted in and sustained by politics rather than economics. According to a report prepared for the Global Subsidies Initiative:

> The failure to reform subsidies fully lies in the failure to appreciate the political economy of subsidy policies. While subsidies are abhorrent to economic analysts and can be a particularly pernicious form of public policy, in most cases subsidies exist because they are rooted in a political

Chapter one | Emmanuel Ojameruaye, in
Jideofor Adibe (Ed.)
The Politics and Economics of Removing Subsidies on Petroleum Products in Nigeria
London & Abuja, Adonis & Abbey Publishers

logic that is often difficult to alter. The interest groups that demand for subsidies are usually well organized, and the provision of a subsidy usually makes these groups even more aware of their interest in sustaining the subsidy policy. Further, the entities that supply subsidies often find political advantage in providing this costly service....Policies that provide subsidies often help leaders achieve that goal (of staying in power) by channeling resources to interest that could affect government survival, such as voting or by donating to their political campaigns... Once a subsidy is created, regardless of its original purpose, interest groups and investments solidify around the existence of the policy and make change difficult (Victor, 2009).

In Nigeria, there has been an unending debate on the "appropriate" pricing of petroleum products since the introduction of fuel subsidies in the 1980s. The debate is driven more by politics than economics. Politics underpins and sustains fuel subsidy in Nigeria. On the one hand, some top government officials and technocrats as well as professional economists argue for the elimination of the subsidies. On the other hand, labor union leaders, civil society activists, politicians and importers of fuel argue for maintaining the subsidies, albeit for different reasons. The opponents of the fuel subsidies claim that the subsidies are doing more harm than good to the economy and that they are unsustainable. They also claim that the subsidies benefit the rich more than the poor and that the funds used to pay for the subsidies can be put to better use for the economy by the government. The proponents of the subsidies believe that the subsidies are necessary to keep prices of goods and services affordable for the poor and that eliminating the subsidies will result in galloping inflation that will adversely affect the poor. They also claim that the government can afford to maintain if it can get rid of the grand corruption that is associated with the subsidy regime.

In the debate over the fuel subsidies, both proponents and opponents have adopted a "war" approach (Ayoola, & Salami, 2010) which is based on the assumption that all public issues are political and politics is about power grab. Politicians seeking to gain power try to "manufacture" the consent of the public and decision-makers by presenting their positions as the "common sense" or dominant one.

Chapter one | Emmanuel Ojameruaye, in
Jideofor Adibe (Ed.)
The Politics and Economics of Removing Subsidies on Petroleum Products in Nigeria
London & Abuja, Adonis & Abbey Publishers

They do so by careful, selective and deceptive use of language. This is why the proponents and opponents of the removal of fuel subsidy have always employed polarizing language or expressions to present their case as the "common sense" one while ignoring the "pure economics" of the issue. For instance, federal government officials who have been arguing for the "removal" of the subsidies have been using expressions to steer people to believe that petroleum products are underpriced and that the subsidies are inimical to the economy. A good example of this is President Jonathan himself. At a meeting with the National Working Committee of the ruling Peoples' Democratic Party held on Friday, October 22, 2011 to sell the fuel subsidy removal plan, the President said that failure to remove the subsidy would lead to the collapse of the Nigerian economy. According to a source at the meeting,

> The President told us that... though he appreciated the pains Nigerians would pass through after the removal, he said it was necessary if we did not want the economy to collapse... the N1.2trn saved from the subsidy removal would be spent on providing social projects for the poor...the planned removal would be followed by a lot of palliatives (*Punch*, October 24, 2011).

Furthermore, at the 17th Nigerian Economic Summit presidential dialogue on Thursday, November 10, 2011, President Jonathan stated that the critics of the fuel subsidy removal policy have turned the issue into full-scale politics with the sole aim of bringing down his government. According to the President,

> The opponents of fuel subsidy removal in their closet saw the wisdom in fuel subsidy removal, but they resorted to politicizing the policy when they saw it as an opportunity to bring the government down (*Nigerian Tribune*, November 11, 2011).

The President also noted that the government had to slow down on the implementation date of fuel subsidies removal in order to ensure that all stakeholders were carried along through mass education and logical reasoning.

Chapter one | Emmanuel Ojameruaye, in
Jideofor Adibe (Ed.)
The Politics and Economics of Removing Subsidies on Petroleum Products in Nigeria
London & Abuja, Adonis & Abbey Publishers

At a symposium organized by the Academic Staff Union of Universities in Ibadan, Adeola Adenikinju, a professor of economics who supports the removal of the subsidy stated that subsidy

> ...has moved from implicit subsidy to explicit subsidy, with the exchange rate used in the importation of oil, depleting the foreign reserves, thus making removal of the subsidy a necessity. If we are to refine locally, all these costs spent on importation will not be necessary (*Daily Sun*, November 16, 2011).

On the other hand, at the same symposium, former petroleum minister Prof. Tam David West, who did not believe fuel was being subsidized stated categorically that subsidy

> "Is a sanctified falsehood, absolute lie...I challenge (President) Jonathan... to prove me wrong...They have killed the refineries through sabotage...You are asking the masses to pay for your inaction". (*Daily Sun*, November 16, 2011).

Another critic of the removal of the subsidy, Afonne (2011), observed that:

> Successive governments at the federal level in Nigeria since 1986, had tinkered in several ways with the price of petroleum products under various guise... government's huge appetite for increase in pump price of petroleum products seems insatiable...Ordinarily, Nigerians would not have lost a minute's sleep on the purported hike in price of fuel, if previous exercises had yielded desired results. Rather, while many people in the country may be feeling the crunch wrought on their already lean pockets by the incessant increase in price of the product, a select group of brief-case-businessmen are smiling to the banks. There is a cartel which has perfected plans to inflict maximum pain on many people in the country... How can one rationalise that in a democracy, where N240 billion was appropriated for 2011 fiscal year by the National Assembly for the so-called fuel subsidy, about N1.3 trillion has already been spent by the Nigeria National Petroleum Corporation, NNPC, and the year has not ended yet. It is obvious that the cartel's fraudulent activities are at work, albeit tacit support from government functionaries.

Chapter one | Emmanuel Ojameruaye, in
Jideofor Adibe (Ed.)
The Politics and Economics of Removing Subsidies on Petroleum Products in Nigeria
London & Abuja, Adonis & Abbey Publishers

Toeing the same line, Ishiekwene (2011) posited that:

> The rulers have been criminal liars … perpetual fraud and deception of the
> people… that petrol in Nigeria is heavily subsidised, and the "fuel subsidy"
> must be removed so government can free itself from subsidising motorists
> and use the money for infrastructure development. The truth of the matter
> is that there has never been any subsidy… a cabal has hijacked the Nigerian
> Petroleum distribution and Marketing process… And so Nigeria has been
> dependent largely on imported petrol and kerosene for a growing
> population.. And who does all the importation of Petroleum products? PDP
> mafias and election rigging funders at all levels of the chain. …There will be
> no end to these cycles of increases and lies of "Subsidy removal" so long as
> the greedy elites… continually depreciates the Naira relative to the Dollar…
> Yes the Nigerian people must resist very vigorously this umpteenth increase
> in fuel prices the name of a bogus fuel subsidy removal.

In a similar vein, in presentation on the fuel subsidy, Onuegbu
(2011) who is the Chairman of the Rivers State Council of the Trade
Union Congress of Nigeria stated that

> …what is required is the removal of corruption and inefficiency in the
> subsidy and downstream sector management, rather than the removal of the
> oil subsidy, as virtually all countries of the world operate one form of
> subsidy or the other for her citizens. … We expect the government to carry
> out a sincere, detailed and comprehensive review of the downstream sector
> with a view to finding and implementing lasting solutions to the industry's
> problems… (*Vanguard*, November 10, 2011).

However, the technocrats in government continued to insist on
eliminating the subsidies. For instance, following the House of
Representatives' Committee's rejection of the President's proposal in
November 2011, the Minister of Finance, Ngozi Okonjo-Iweala stated
that:

> Subsidy does not get to the poor; the middle and upper classes are the real
> beneficiaries. It is clearly unsustainable… Evidence shows that the price of
> fuel in Nigeria is below both the African and international average… We will
> be better off using the amount spent on subsidy to target poorer groups and
> big infrastructure projects… The subsidy causes distortions that result in

Chapter one | Emmanuel Ojameruaye, in
Jideofor Adibe (Ed.)
The Politics and Economics of Removing Subsidies on Petroleum Products in Nigeria
London & Abuja, Adonis & Abbey Publishers

huge economic costs such as rent-seeking behaviour and smuggling...
...Subsidy is a major fiscal and financial burden on the nation. ...
Deregulation implies limited intervention by government; it allows for
better regulation and transparency; allows for free operation activities in the
sector; attracts new investors into the market and it increases competition
and promotes overall higher productivity; reduces scarcity by ensuring
adequate supply of petroleum products; and similar success story to the
telecommunication sector (*Thisday Live,* December 2, 2011).

The Minister also stated that some social safety nets that the
government will implement if the subsidy is removed include the
following: a) Launching of Subsidy Reinvestment and Empowerment
Programme (SURE); b) Maternal and child health services; c) Public
works/youth employment programme; d) Urban mass transit scheme;
e) Vocational training schemes; and f) High-profile infrastructure
projects: Roads and rail; water resources, power; refineries (with
private sector). She insisted that

> Structures have been developed to guarantee adequate oversight,
> accountability and implementation of the various projects...To ensure
> effectiveness, efficiency and delivery, high powered committee of eminent
> Nigerians to monitor revenue proceeds and proper implementation and use
> of the amount saved. Members with proven integrity will be drawn from the
> Nigerian youth, women groups and civil society organisations. (*The Nation,*
> December 5, 2011).

But many Nigerians doubted the Minister's assurances because
they see her as a World Bank economist and remembered the
Structural Adjustment Program (SAP) which the IMF and the World
Bank foisted on Nigeria in 1986 and which failed to deliver on its
promises. Many Nigerians also remembered that the Minister was
also the Minister of Finance when President Obasanjo started the
process of deregulation of petroleum product prices in 2002 which
also led to increases in fuel prices without the provision of the
promised social safety nets.

During a televised Town Hall Meeting organized by the
Newspaper Proprietors Association of Nigeria (NPAN) on December

Chapter one | Emmanuel Ojameruaye, in
Jideofor Adibe (Ed.)
The Politics and Economics of Removing Subsidies on Petroleum Products in Nigeria
London & Abuja, Adonis & Abbey Publishers

20, 2011 which featured three top government technocrats (Petroleum Minister Diezani Alison Madueke, Finance Minister Ngozi Okonjo-Iweala and Central Bank Governor Sanusi Lamido Sanusi) and some opponents of the subsidy removal, . Ngozi Okonjo-Iweala further reiterated government's case against subsidy as follows:

> The current subsidy regime does not benefit the poor in Nigeria. It is the better-offs in the society who benefit the more from it…'Between January and October this year we've spent over N1.3 trillion being borrowed and it is used to subsidized petroleum products consumed mostly by the upper and middle class who own the big cars, SUVs, jeeps, and not the poor who drive in bicycles or motorcycles. We have to put an end to subsidy; it is not good for the economy" (*The Nigerian Voice*, December 26, 2011.

Corroborating the Finance Minister at the Town Hall meeting, the Governor of the Central Bank of Nigeria stated that:

> I want an end to it soon…We are borrowing from our children to sustain a lifestyle that favours only the elites and not the poor…'This government can continue to pay this subsidy, but it will be catastrophic for the next government if this government does not stop the payment of this type of money on subsidizing fuel…This subsidy creates an opportunity for corruption in the system. ..There is so much opportunity in the import business for fraud because of rent seekers. I am a proponent of the removal of rent. There is a danger ahead we have to avoid (*ibid*)

However, an opponent of the subsidy removal, the Vice President of the Nigeria Labour Congress, Isa Aremu, reiterated the stand of the labour unions as follows:

> Our fear is that if we allow the government to remove subsidies, it is not going to be a win-win situation for government and citizens. Removing subsidy will worsen the poverty situation in this country because inflation will go up. And we are not prepared for this…Nigerians will resist attempts by this government to force citizens to pay higher prices for petroleum products imported from outside the country while we could have paid cheaper if we had our refineries working (*ibid*)

Chapter one | Emmanuel Ojameruaye, in
Jideofor Adibe (Ed.)
The Politics and Economics of Removing Subsidies on Petroleum Products in Nigeria
London & Abuja, Adonis & Abbey Publishers

In spite of the opposition of the labour unions and leaders of some civil society organizations, President Jonathan went ahead to remove the subsidy on January 1, 2012. However, the "removal" was short-lived because of violent protest which forced the President to bring down the price of petrol from N141 per litre (the "market" price") to N97 per litre on January 16, 2012. The price of petrol has remained at that level since then while the official price of kerosene has also remained at N50 per litre but the price of diesel has remained deregulated. Thus the subsidy regime has remained and the government paid over N2 trillion in subsidies to NNPC and petroleum products importers and marketers in 2012!

Thus, while the government appears committed to reducing or removing fuel subsidy, many people and trade unions do not buy the argument of the government or trust the government. At times, for political reasons, the government appears wary or ambivalent on whether and when it will eliminate the fuel subsidy. For instance, the National Assembly appears unwilling to approve or entertain a proposal from the President to remove or reduce the subsidy due to the political and social fallouts. This is probably why the President bypassed the National Assembly on January 1, 2012 during the botched subsidy removal attempt. In fact, before the botched attempt, there were rumors that the government had bribed members of the National Assembly with N2 billion to approve the removal of the subsidy, but it was denied. According to a government's spokesman,

> Whatever is the decision of the National Assembly on the withdrawal of fuel subsidy is advisory and Section 16(2) of the Constitution gives the government the power to remove the subsidy because it is purely economic matter even if the National Assembly does not approve it (Leadership, December 5, 2011).

This was probably why the President ignored the National Assembly and voices of the people and went ahead to announce the ill-fated removal of the subsidy on January 1, 2012.

Chapter one | Emmanuel Ojameruaye, in
Jideofor Adibe (Ed.)
The Politics and Economics of Removing Subsidies on Petroleum Products in Nigeria
London & Abuja, Adonis & Abbey Publishers

It is not clear if and when the government will attempt to remove or reduce the subsidy again, but as the country approaches a presidential election in 2015 the decision will be increasingly difficult to make. Meanwhile, the government continues to send conflicting signals of its intention on the subsidy issue (Ojameruaye, 2012b). However, it is almost clear to all now that the corrupt fuel subsidy regime is unsustainable and uneconomical. Sooner or later, the government must adopt a pricing approach that will put to an end to the recurring fuel subsidy controversy and episodic "removal" of subsidy.

References

Adenikinju, A (2009), "Energy pricing and subsidy reform in Nigeria". A presentation at the Global Forum on Trade and Climate Change. OECD Centre, 9-10 June 2009. Available at: http://www.oecd.org/dataoecd/58/61/42987402.pdf (Accessed: 5 April, 2013)

Afonne, E. (2011), "Politics of Oil Subsidy: The Cartel's Fraudulent Acts". Nigerian NewsWorld, October 24, 2011. Available at: www.nigeriannewsworld.com/content/politics-oil-subsidy-cartel%E2%80%99s-fraudulent-acts-october-24-2011. (Accessed: 15 May 2013)

Ayoola, K.A. & Salami, O. (2010), "The 'War' of appropriate Pricing of Petroleum Products: The Discourse of Nigeria's Reform Agenda. *Linguistik Online*, 42, 2/2010. http://www.linguistik-online.de/42_1 0/salamiAyoola.html_(Accessed: 15 May 2013)

Ishiekwene, T. (2011), "Why Nigerians Must Resist The Next Fuel Subsidy Withdrawal". Available at http://saharareporters.com/arti cle/why-nigerians-must-resist-next-%E2%80%9Cfuel-subsidy%E2%80%9D-withdrawal (Accessed: 20 May 2013)

Kalu, I.K. (2012), Full Report of the Refineries Special Task Force headed by Dr. Kalu Idika Kalu. Available at www.elombah.com (Accessed: 22 May 2013)

Chapter one | Emmanuel Ojameruaye, in
Jideofor Adibe (Ed.)
The Politics and Economics of Removing Subsidies on Petroleum Products in Nigeria
London & Abuja, Adonis & Abbey Publishers

Makwe, I.A. (2006), "A critique of the Nigerian Petroleum Products Pricing Regulatory Agency (PPPRA) Pricing Template and Cost Recovery Analysis". *Oil, Gas and Energy Law* (OGEL), Vol. 3, September.

Nuhu-Koko, A. (2008), "Fuel subsidy scandal and the impending subsidy removal". *Nigeria Energy Intelligence.* 10 November.

Ojameruaye, E.O. (2011), the political economy of the removal of petroleum products subsidy in Nigeria. Parts I and II. Available at:___http://chatafrik.com/articles/economy/item/312-the-political-economy-of-the-removal-of-petroleum-products-subsidy-in-nigeria-part-i-the-politics.html

Ojameruaye, E.O. (2012a), "The Petroleum Subsidy Probe Report: Some Technical Concerns and Weaknesses". Available at: http://chatafrik.com/articles/nigerian-affairs/item/929-the-petroleum-subsidy-probe-report-some-technical-concerns-and-weaknesses.html

Ojameruaye, E.O. (2012b), "Removing Petroleum Products Subsidy in Nigeria, To be, or Not to be". Available at: http://chatafrik.com/articles/nigerian-affairs/item/1304-removing-petroleum-products-subsidy-in-nigeria-to-be-or-not-to-be?Html

Onumah, C. (2012), "The fuel subsidy conundrum". Available at: http://premiumtimesng.com/opinion/108259-the-fuel-subsidy-conundrum-part-1-by-chido-onumah.html. (Accessed 6 April, 2013)

Peterson, W.W.F. (2009), A Billion Dollars a Day: The Economics and Politics of Agricultural Subsidies. Wiley-Blackwell.

U4 AntiCorruption Resource Centre (2009), Reforming corruption out of Nigerian oil. U4Brief No. 2, February, 2009. Available at: www.u4.no/themes/nrm_(Accessed 12 April, 2013)

Victor, D. (2009), "Untold Billions: Fossil-Fuel Subsidies, Their Impacts and the Path to Reform. The Politics of Fuel Subsidies". A paper prepared for the Global Subsidies Initiative (GSI). The International Institute for Sustainable Development (IISD). Geneva, Switzerland. October 2009.

Chapter one | Emmanuel Ojameruaye, in
Jideofor Adibe (Ed.)
The Politics and Economics of Removing Subsidies on Petroleum Products in Nigeria
London & Abuja, Adonis & Abbey Publishers

46

World Trade Organization (2006), "The economics of subsidies" in 2006 World Trade Report: Exploring the links between subsidies, trade and the WTO. Available at: http://www.wto.org/english/res _e/booksp_e/anrep_e/world_trade_report06_e.pdf   (Accessed: 15 April, 2013)

Chapter one | Emmanuel Ojameruaye, in
Jideofor Adibe (Ed.)
The Politics and Economics of Removing Subsidies on Petroleum Products in Nigeria
London & Abuja, Adonis & Abbey Publishers

CHAPTER TWO

Petroleum Subsidy Debate in Nigeria: Issues and Policy Options

Ogujiuba Kanayo

Abstract

Nigeria has one of the highest poverty rates globally; and that rate may not abate in the medium-term. Owing to the critical role of petroleum products in the daily life of Nigerians, it is believed by many that the multiplier effects of petroleum subsidy removal will not only push more people below the poverty line, they will also wipe off jobs, worsen crime rates and stifle efforts by the country to meet the United Nations Millennium Development Goals. Most policy analysts fear that the so-called subsidy removal on petrol may have the same impact as the "deregulation" of diesel under President Olusegun Obasanjo which raised the operating cost of manufacturers that hastened the exit of numerous companies from Nigeria. Howbeit, the removal of subsidy would lead to an increase in inputs cost, which would be higher than the increase in selling prices of firms, and the lack of sequencing the proceeds based on needs would leave the poverty level at status-quo. Key sectors of the economy which may experience increased nominal output as a result of petroleum subsidy removal are the refined petroleum products, which provide income for an extremely low number of households. Nonetheless, policy responses could either be positive or negative depending on the nature and timing. An expansionary policy of spending the proceeds from subsidy removal would likely favour rural households and disfavor urban households, because urban households earn most of their incomes from inputs-intensive sectors while rural households do not. Furthermore, an expansionary policy stance would

Chapter Two | Ogujiuba Kanayo, in
Jideofor Adibe (Ed.)
The Politics and Economics of Removing Subsidies on Petroleum Products in Nigeria
London & Abuja, Adonis & Abbey Publishers

consequently fuel inflation and worsen urban income while it improves rural income, as output prices rise, generally. Thus, a non-inflationary expansionary policy which increases transfers to households would have the least poverty effect. Nonetheless, inflation resulting from subsidy removal can be reduced considerably with a conservative fiscal policy response. This chapter recommends that (1) the relative effects of the petroleum subsidy removal on different socioeconomic groups should be explored before implementation, to avoid ambiguous responses and (2) that incorporating information on the dynamics and heterogeneity of household income sources would be beneficial to the government in the sequencing of the proceeds of the subsidy removal.

1.0 Introduction

Nigeria is rich but its people are poor (World Bank, 1996).This irony has made it imperative to assess the poverty implications of the government's activities. A greater urgency should be brought into this issue as the population of poor people is almost steadily growing: between 1980 and 2010 the percentage of poor people (living below the poverty line) increased from about 27% [18 million] to over 72% [85 million] of the population (NBS 2011). Poverty is caused and affected by both microeconomic and macroeconomic as well as socio cultural factors. This is further compounded by the possible impact of petroleum products pricing and subsidizing which has been the focus of much controversy with regard to poverty dynamics in Nigeria. The issue has dominated economic debate in recent times even before the subsidy was recently substantially withdrawn by the federal government in January, 2012. The partial removal of fuel subsidy by the Federal Government as a deliberate policy aimed at conserving and maximizing the oil wealth of the world's sixth highest petroleum producer/exporter has attracted national and international discourse since the policy took place. In addition to hike in tuition fees in higher educational institutions and chemical fertilizers, as a result of the

Chapter Two | Ogujiuba Kanayo, in
Jideofor Adibe (Ed.)
The Politics and Economics of Removing Subsidies on Petroleum Products in Nigeria
London & Abuja, Adonis & Abbey Publishers

subsidy removal, prices of all basic petroleum products such as premium motor spirit or petrol, diesel, aviation and household kerosene jerked up to reflect the interplay of 'free' market forces of supply and demand. It is important to note that the debate on the removal of subsidies on petroleum products in Nigeria dates back to over a decade. The issue of removal of subsidy from petroleum products consumed in Nigeria is not new; for about 30 years since 1982 when President Shehu Shagari introduced the first set of austerity measures to address the then worsening economic situation faced by the Nigerian economy under his leadership deregulation and removal of all subsidies from domestic prices of consumer products and services have been on the front burner and were two of the kernels of the IMF/World Bank-supported structural adjustment Programme (SAP) which was in place between 1986 and 1991 (Abubakar, 2011).

Over these years, domestic prices for fossil fuels rose from an average of 50 kobo per litre for petrol, diesel, aviation kerosene and household kerosene to the present pricing regime that pegged the retail price of petrol per litre at N97, diesel at N115 and Kerosene at N50 as Federal Government approved prices for these products. However, there are two main reasons government is putting forward in support of the deregulation of the petroleum downstream sector include; firstly, the need to free more funds to tackle the huge infrastructural deficit the country is facing. Secondly, the need to free the market in order to attract private investments especially in building refineries to meet at least the local consumption demand and to create jobs. However, these reasons are subjects of debate because deregulation is an essential but not sufficient condition to guarantee private investments.

This chapter examines the debate on the effects of subsidy on petroleum products, the linkage between petroleum products' prices and poverty as well as recommends policy options.

Chapter Two | Ogujiuba Kanayo, in
Jideofor Adibe (Ed.)
The Politics and Economics of Removing Subsidies on Petroleum Products in Nigeria
London & Abuja, Adonis & Abbey Publishers

2.0   The Debate: Petroleum Subsidy

There are two types of subsidies generally referred to in the literature: explicit and implicit subsidy. Explicit subsidy is the difference between production cost and selling price. Implicit subsidy on the other hand refers to the difference between the opportunity cost of a wasting asset and the present selling price. Actually this is what is usually meant when subsidy of oil products is involved (Adenikinju 2000, IMF 2003).   Implicit subsidy is important because of the implications for efficiency. For pricing in the sub sector to be efficient, prices should normally be equal to the Marginal Opportunity cost. For the petroleum sub sector, this is the border or international price of the product (Adenikinju, 2001; Hossain, 2003). This is necessary so as to compensate future generations for the irreversible extraction of the product so that a foundation for continued growth even when the petroleum resources are exhausted is laid for future generations. In the recent past, the domestic prices of petroleum products in Nigeria were much lower than what obtains in the neighbouring countries, which has led to a thriving smuggling business. This difference is partly because the crude oil for producing products for domestic consumption is sold to the local refineries at a lower price per barrel. This therefore brings down the cost of production.

Nonetheless, the substantial removal of government subsidy on petroleum products in January, 2012 has generated a lot of controversies in the country. The debate on the benefits or otherwise of the policy has given birth to two major schools of thought: (i) The government and those who believe in Free Market Economy are supporting the policy, (ii) while the Organized Labour and few Petroleum Products Marketers are advocating for the sustenance of subsidy on petroleum products (Abiola, 2010). The federal government posits removal of petroleum subsidy was necessary to save the economy from collapsing; and is arguing that this policy measure was capable of enhancing the economic fortunes of the country. Furthermore, government argues that the oil subsidy which

Chapter Two | Ogujiuba Kanayo, in
Jideofor Adibe (Ed.)
The Politics and Economics of Removing Subsidies on Petroleum Products in Nigeria
London & Abuja, Adonis & Abbey Publishers

hitherto was only beneficial to a few Nigerians would now be diversified to improve the lot of a greater majority of the less-privileged Nigerians via the provision of infrastructures, improvement in education, health, power, water resources and agriculture.

Mukhtar (2012) submitted that there have been lots of discussions, claims and counter claims, threats and counter threats, etc., on the deregulation of the downstream petroleum sector. This came to a peak in January, 2012 with the government's announcement that the deregulation had started and the immediate effect of this was the substantial removal of the subsidy on petroleum products[2]. First of all, there are many Nigerians who believe that there is no petrol subsidy in Nigeria instead it is even over priced at N65/litre. According to an article, "The Truth about Oil Subsidy" by Ganiyat Gani-Fawehnmi (wife of late Chief Gani Fawehnmi),

> ...the truth is that there was never an oil subsidy; there has never been an oil subsidy and today there is no oil subsidy in the pricing of petrol per liter in Nigeria. The causes of our present oil chaos are not the issue of oil subsidy but: High level of corruption in all strata of governance in all parts of Nigeria: Massive and unchecked stealing by our leaders, their cohorts and cronies in public and private sectors of the Nigerian economy over the decades: Open and deceptive mismanagement of our resources including public funds: Mindless and mind-boggling lavish projects specifically designed as conduit pipes to siphon the people's common wealth into private pockets at the expense of the needs and cares of the suffering Nigerian masses: and Unceasing and measured astronomical devaluation of the Nigerian currency, a result of gross mis-governance of the country in all facets of human activities(cited in Adejumo, 2012).

Adejumo, (2012) also reported the result of a 2010 research conducted by Strategic Union of Professionals for the Advancement of Nigeria (SUPA) which concluded that the government makes a

---

[2] Subsidy paid by the government is the difference between the landing cost of imported fuel and government fixed pump price. (N141 – N65 = N75 per litre for PMS and N148.98 – N50 = N98.98 for HHK).

Chapter Two | Ogujiuba Kanayo, in
Jideofor Adibe (Ed.)
The Politics and Economics of Removing Subsidies on Petroleum Products in Nigeria
London & Abuja, Adonis & Abbey Publishers

53

profit of N33.50 /Litre on PMS (Petrol) at the price of 65 N/Litre. This translates to a very high 106% profit per litre. In addition the government benefits from royalties, taxes and fees which were not factored in this simplified analysis. The research indicates, that when factored, the actual crude cost per barrel to government is significantly less and its profit correspondingly higher. In another research, a litre of petrol should cost between N35-N45, and argued that the 445,000 barrels per day, which is meant for domestic consumption, should not be for sell to Nigerians (Guardian Newspaper, Dec. 28, 2011). It should be free of charge so only the cost of finding/development, production/storage /transportation, refining, refined products pipeline transportation, distribution margins (retailers, transporters, dealers, bridging funds, administrative charges etc.) should be considered. On the contrary, the governments' position is that the landing cost of a litre of petrol is about N123/litre based on an average crude oil price of about US$ 114/barrel and the distribution margins of about N16/litre. This means, according to the government, that the supply cost is N139/litre and the government is paying N74/litre while people only pay N65/litre (Public Presentation by the Finance Minister, Dec. 6, 2011). Most people are opposed to the removal of fuel subsidy because they are not sure that such funds would be accounted for; and many have asserted that the Federal Government should have extended the period of awareness and sensitization to enable all stakeholders to actually appreciate government's good intentions and thereafter in a very calm atmosphere start a graduated removal of the oil subsidy with little or no stress (Nwadialo, 2012). The staunch opposition and the fears of the leaderships of the organized labour and trade unions is appreciated from the literal interpretation of deregulation or the face value of removal of subsidized pump price of imported petroleum products[3]..

---

[3] Their apparent innocence and efforts were probably aimed at protecting the common man in the street from an untold hardship of being exploited and exposed

Chapter Two | Ogujiuba Kanayo, in
Jideofor Adibe (Ed.)
The Politics and Economics of Removing Subsidies on Petroleum Products in Nigeria
London & Abuja, Adonis & Abbey Publishers

Who is really right in this debate? First and foremost, subsidy is the money paid by government or an organization to make prices lower or reduce the cost of producing goods. Deregulation on the other hand, means the removal of government rules and controls from some types of business activity. The primary essence and idea of subsidy in effect, is targeted at benefiting the masses, the very poor in our system but, in Nigeria's context, controversies have trailed the supposed impact of this policy. The benefit incidence on the poor, have been very abysmal; furthermore, empirical evidence from several studies and surveys have shown that the bulk of the limited resources of the country are enjoyed by very few members of the ruling political, bureaucratic and business class (the elites) at the expense of the very many poor members of the society (Nwachukwu, 2011). This supposes that the huge amount of money pulled by the Federal Government to stabilize the domestic pump price of imported petroleum products in the face of high landing costs (the amount between the landing cost of the imported products and the agreed ex-depot price by the NNPC) could be better utilized within the country if the country can comfortably sustain its daily consumptions of refined petroleum products through local refineries. Howbeit, the Nigerian nation suffers double jeopardy economically; huge amounts are wasted annually on the subsidy (without achieving the primary aim of poverty reduction for the poor masses) and at the same time, the country 'exports labor' to the countries of origin of those refined petroleum products with her hard earned foreign exchange (instead of creating jobs locally through local refining).

Furthermore, the federal government has maintained that the subsidy cost is between N1.3 - N1.5 trillion annually, which represents about 30% of its total expenditure though, some are questioning this figure based on the point that the actual daily consumption of petrol is probably not known (Gamji Website, Jan. 8,

---

to high transportation costs and its adverse effects on the prices of other commodities

Chapter Two | Ogujiuba Kanayo, in
Jideofor Adibe (Ed.)
The Politics and Economics of Removing Subsidies on Petroleum Products in Nigeria
London & Abuja, Adonis & Abbey Publishers

2012)[4]. The Federal Government maintains that savings from subsidy removal will reduce borrowing; preserve reserves; create revenue, support free market operations and open the sector for new private investments. Also, the government argues that the subsidy only benefits the rich and the middle class, the petroleum marketers, smugglers, the neighboring countries and that Nigeria is not as rich as the other oil producing countries that have subsidized fuel (Mukhtar, 2012) Nonetheless, majority of Nigerians in the urban areas rely on public transport and in rural areas, goods are mostly transported using petrol-engine vehicles not diesel-engine vehicles (NBS 2011). Apart from the transportation sector, millions of Nigerians use petrol-based generators to run their small businesses for their daily income.[5]. Furthermore, the point that Nigeria is not rich compared to some oil producing countries in terms of availability of infrastructure, crude oil production volume per population and reserves (both resources & money) is true[6]. But it is also true that majority of Nigerians are too poor (NBS, 2011) to afford any further squeezing of their income.

3.0 Petroleum Prices - Poverty Linkages

The return to civilian administration in 1999 ushered in President Obasanjo. While in office (1999 - 2007), President Obasanjo raised the

---

[4] Government should make extra effort to verify the actual consumption demand, which may bring some savings.

[5] The rich people, medium and large-scale businesses (owned by the rich people) use bigger diesel-engine generators so virtually they are not affected by the current removal because diesel has no subsidy (see Fuel Subsidy Removal Debate: The Way Forward by Dr. Bello Mukhtar in http://www.gamji.com/article9000/NEWS9588.htm retrieved 02/09/2013.)

[6] For example, statistics for availability of Infrastructure shows that US ranks 14 with 5.81 scores, Saudi Arabia 26 with 5.23, Libya 81 with 3.56 compared to Nigeria that ranks 130 with 2.28 scores (see WEF Global Competitiveness Report 2012-2013). In crude oil production, recent information reveals that Russia produces the highest in the World with 10.7 Million (bbl/day), Saudi Arabia with 9.5M (bbl/day), US 9.0M(bbl/day), and Iran 4.2M(bbl/day), compared to Nigeria production with 1.9M(bbl/day) in 2013 (see http://www.opec.org/opec_web/en/).

Chapter Two | Ogujiuba Kanayo, in
Jideofor Adibe (Ed.)
The Politics and Economics of Removing Subsidies on Petroleum Products in Nigeria
London & Abuja, Adonis & Abbey Publishers

prices for domestic petroleum products 13 times. The former president met prices pegged at N11 per litre and by the time he vacated office, the price of petrol per litre was N75. It was late President Umaru Musa Yar'adua[7] who reduced the price of PMS to N65 per litre and the current pricing regime of N97 per litre was introduced by President Goodluck Jonathan in January, 2012.

Table 1 below, shows the growth rate of the price of Premium Motor Spirit (PMS) in Nigeria vis-à-vis poverty and unemployment

Table 1: Growth Rate of the Price of Premium Motor Spirit (PMS), Poverty & Unemployment in Nigeria from 1990 to 2012

Year	Price of Premium Motor Spirit (PMS) Naira	% Increase (Decrease)	Poverty Rates	Unemployment Rates	Government in Power
1990	0.7	-	44.0	3.5	Military /Babangida
1991	0.7	0.0	43.5	3.1	Military/ Babangida
1992	5.00	614.3	47.7	3.5	Military/ Babangida
1993	3.25	(35)	49.0	3.4	Military/ Babangida
1994	11.00	238.5	54.1	3.2	Military/Abacha
1995	11.00	0.0	60.0	1.9	Military/ Abacha
1996	11.00	0.0	65.6	2.8	Military/ Abacha
1997	15.00	36.4	65.5	3.4	Military/ Abacha
1998	15.00	0.0	69.6	3.5	Military/ Abacha
1999	20.00	33.3	72.0	17.5	Civilian/Obasanjo
2000	22.00	10	74.0	13.1	Civilian/ Obasanjo
2001	26.00	18.2	83.1	13.6	Civilian/ Obasanjo
2002	30.00	15.4	88.0	12.6	Civilian/ Obasanjo
2003	40.00	33.3	79.3	14.8	Civilian/ Obasanjo
2004	49.00	22.5	54.7	13.4	Civilian/ Obasanjo

---

[7] Successor to President Obasanjo.

Chapter Two | Ogujiuba Kanayo, in
Jideofor Adibe (Ed.)
The Politics and Economics of Removing Subsidies on Petroleum Products in Nigeria
London & Abuja, Adonis & Abbey Publishers

2005	52.00	6.1	76.3	11.9	Civilian/ Obasanjo
2006	64.50	24.0	74.6	12.3	Civilian/ Obasanjo
2007	75.00	1.6	34.1	12.7	Civilian/ Obasanjo
2008	75.00	0.0	59.9	14.9	Civilian/ Yar'adua
2009	65.00	(13.3)	61.2	19.7	Civilian/ Yar'adua
2010	65.00	0.0	60.2	21.1	Civilian/ Yar'adua
2011	65.00	0.0	71.1	23.9	Civilian/ Jonathan
2012	97.00	49.2	-	-	Civilian/ Jonathan

Source: *Author's Computation*

Subsidies are supposed to be granted to aid the poor, stabilize prices, and promote economic growth, among others. Nonetheless, if subsidies on petroleum products are removed, the money so saved can be channeled to provide well-developed infrastructure (in particular, power supply and transportation) and human capital development (education, agriculture and healthcare), security and stable political environment which would benefit both the poor and the rich. Besides, subsidy removal will empower the private sector in order to facilitate and expand production, create jobs and consumer demands, attract foreign direct investments, which will help the country to move from an import dependent economy to an export oriented one. Figure 1 below shows the supposed linkage between the removal of subsidies on petroleum products and poverty.

Chapter Two | Ogujiuba Kanayo, in
Jideofor Adibe (Ed.)
The Politics and Economics of Removing Subsidies on Petroleum Products in Nigeria
London & Abuja, Adonis & Abbey Publishers

Fig: 1. Petroleum – Poverty Linkages

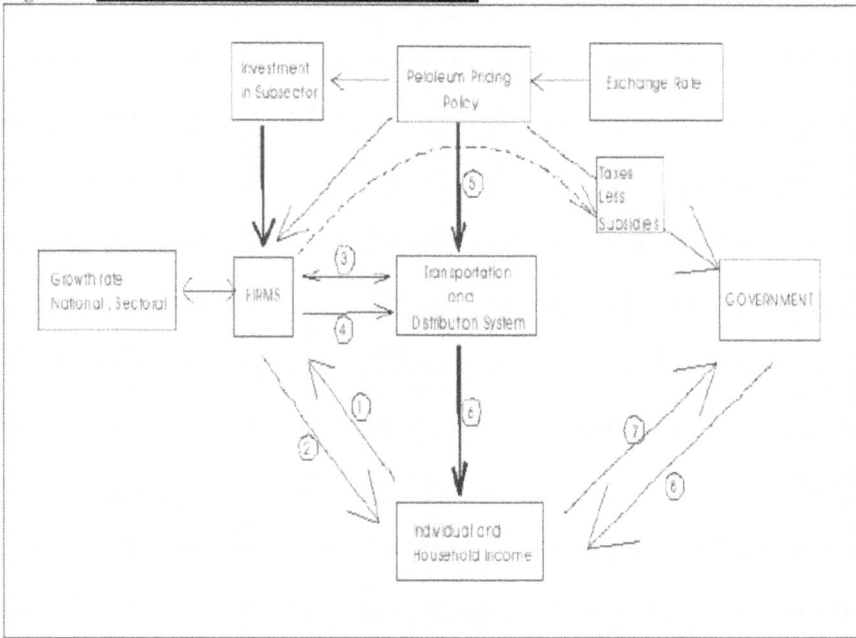

Source: Nwafor M; Ogujiuba K & Asogwa R (2006): Does Subsidy Removal Hurt the Poor; SISERA Working Paper Series

The above diagram shows three major channels that account for the effects of petroleum prices on poverty:

a) Impacts on firms;
b) Impact on and off the distribution and transport system; and
c) Impact on government income and expenditure.

A sizeable number of firms in Nigeria depend on petroleum powered generating sets for their energy supply as electricity supply is grossly inadequate and/or unreliable. Thus, as firms' energy bill increases; the cost of intermediate inputs also increases as a result of increased cost of transportation of individuals and goods; furthermore, increases in private investment in the sub sector are

Chapter Two | Ogujiuba Kanayo, in
Jideofor Adibe (Ed.)
The Politics and Economics of Removing Subsidies on Petroleum Products in Nigeria
London & Abuja, Adonis & Abbey Publishers

expected as it becomes more attractive. The effects on the firm translate into higher cost of doing business which will affect the cost of intermediate and finished goods. This will increase the cost of doing business as well as the output level and profitability of firms as they operate within their budget constraint. The effect on the different firms' behaviour causes changes in the growth rate of the different sectors and GDP.

Also, another pathway through which the change in prices will affect households is the transportation and distribution network as depicted in the diagram. This network is powered by Petroleum products and consequently it has strong inter-sectoral linkages with the sub sector. The existence of a strong linkage is supported by the findings of Adenikinju (2000): following a subsidy reduction the highest increases in prices are in the energy sector followed by the transport sector. As a result of this linkage, increases in the prices of petroleum products lead to increases in passenger and goods transportation cost[8].

In addition, the other pathway through which the changes in prices affect the household is through impact on government revenue and expenditure. Subsidy removal increases government revenue and therefore possible increased government expenditure. However subsidy removal also leads to lower revenue because of the slowdown in growth rate which the increase in prices will cause. It has been noted that the increases could lead to a slowdown in growth of national income (Abel and Bernanke, 1992, Adenikinju 2000) . Adenikinju (2000) reports a fall in real GDP due to a reduction in economic activities. This is related to limits imposed on firms by their budget constraints. Abel and Bernanke report that increases in energy prices in the US due to external oil price shocks , led to reduced energy consumption and reduced output at given levels of capital

---

[8] Ultimately, the increase in transportation cost results in further increase in the cost of intermediate and finished goods in addition to increases attributable to the cost of energy.

Chapter Two | Ogujiuba Kanayo, in
Jideofor Adibe (Ed.)
The Politics and Economics of Removing Subsidies on Petroleum Products in Nigeria
London & Abuja, Adonis & Abbey Publishers

and labour . Generally, as growth of national income is affected, government tax revenue will also be affected due to changes in the tax base. On the expenditure side, Government spending on transfers could increase due to greater revenue availability, which will ultimately improve household welfare[9].

Nigeria still has one of the highest poverty rates globally despite the incessant increases in the prices of petroleum products through subsidy removal and deregulation since 1990 and this rate may not abate in the medium term. Owing to the critical role of petroleum products in the daily life of Nigerians, it is believed by many Nigerians that the multiplier effects of petroleum subsidy removal will not only push more people below the poverty line, but will also wipe off jobs, worsen crime rate and stifle efforts by the country to meet the Millennium Development Goals of reducing poverty by half by the year 2015. Social commentators argue that the so-called subsidy removal on petrol may have the same impact as the deregulation of diesel under President Olusegun Obasanjo, which raised the operating cost of manufacturing and hastened the exit of numerous companies from Nigeria. Furthermore, the removal of subsidy could lead to an increase in input costs, which would be higher than the increase in selling prices of firms and the lack of sequencing the proceeds based on needs would leave the poverty level at status-quo. Furthermore, an expansionary policy stance as a result of the savings from subsidy removal could fuel inflation and worsen urban income, improves rural income and ultimately promotes income inequality. This is because an expansionary policy of spending the proceeds from subsidy removal would likely favour rural households and disfavor urban households. The reason is that urban households earn most of their incomes from inputs-intensive sectors while rural households do not.

---

[9] If this is done without substantially increasing government expenditure it will have the advantage of not putting further pressure on aggregate demand and causing possible demand pull inflation.

Chapter Two	Ogujiuba Kanayo, in
	Jideofor Adibe (Ed.)
	The Politics and Economics of Removing Subsidies on Petroleum Products in Nigeria
	London & Abuja, Adonis & Abbey Publishers

Neo-liberalism maintains that the best way to bring down the price of any commodity is to flood the market with the product. Removing subsidy from petroleum products would continue to increase the prices of petroleum products and other products as well. The argument that deregulation and subsidy removal would encourage private sector to invest in refinery in Nigeria may not stand the test of time. This is because even after the deregulation, the private companies can still refuse to build refineries within the country but rather continue to import refined products. This is because no private company would want to enter into an agreement with a government and after few years, the agreement is put on-hold because of change in administration. Secondly, no private company will want to invest millions of dollars to build refineries, only for the pipelines to be blown-up, employees taken hostage or even worst, killed[10]. Since the recent deregulation of 2012, instability in the sector is very much noticeable. It is argued that availability of infrastructure, security and stable political environment, are first-order conditions for deregulation policy to succeed.

4.0 Issues and Policy Options

Nigeria is among the world's leading oil and gas producing countries. Unfortunately, it is also a nation steeped in paradox; an energy-rich country wracked by fuel and power shortages that are preventing infrastructural progress and stifling economic development. Nigeria has massive gas reserves which are flared as oil is produced, on daily basis. This uncontrolled gas flaring in addition to loss of revenue to the country has altered the local climate of the immediate environment where oil is produced. Nigeria is one of the countries where the price of its domestic oil has been on the increase since

---

[10] Within the last two decades, a number of factories in Nigeria, especially in the textile and agro allied industries have closed production because of instability in the polity, which led to loss of jobs and negatively affected government's revenue. .

Chapter Two | Ogujiuba Kanayo, in
Jideofor Adibe (Ed.)
The Politics and Economics of Removing Subsidies on Petroleum Products in Nigeria
London & Abuja, Adonis & Abbey Publishers

1970s. This is in spite of the fact that the country, in addition to having four major oil refineries, also imports refined products to satisfy its domestic consumption. The crisis in the downstream segment of the Nigerian petroleum sector has therefore been a major concern to most people in Nigeria. Nonetheless, the constant fuel pump price increase in the country has been traced to the inefficiencies of the nation's refineries in addition to the sabotage from bunkerers, oil spillages and attitude of some marketers (See Ogunbodede et al 2010). The effect of this on the nation's economy is constant fuel supply disruptions leading to both economic and environmental problems.

Ogunbodede et al (2010) submitted that in Nigeria, there is a multiple negative effect of incessant increase in the price of crude oil on the economy, because whatever happens in the oil sector affects all other sectors of the economy and by implication, it affects the macro-economic policies of the country. Premium motor spirit (petrol) is needed to power automobiles which farmers and non-use. Similarly, petrol is needed to power generating sets in the country because electrical energy supply from Power Holding Company of Nigeria (PHCN) is not regular and at best epileptic.

Evidence- Based Analogy suggests that the Federal Government and the other tiers of government should continually seek for meaningful dialogue with the organized labour and other stakeholders before implementing full removal of fuel subsidy and further deregulation of the petroleum downstream sector. Thus, there is the need to carry out comprehensive maintenance of the refineries in Nigeria so that they can function at full capacity to refine the petroleum products needed by the country with a view to reducing importation of petroleum products. If these government-owned refineries are too old to function at full capacity after the comprehensive maintenance, it may become necessary to involve communities in setting up small size refineries, to be designed, fabricated and operated by unemployed Nigerian graduates. All the

Chapter Two | Ogujiuba Kanayo, in
Jideofor Adibe (Ed.)
The Politics and Economics of Removing Subsidies on Petroleum Products in Nigeria
London & Abuja, Adonis & Abbey Publishers

government needs to do is to regulate the practice by establishing rules and guidelines and supplying the refineries with crude oil at current price value in the international crude oil market[11]. Crude oil is very easy to move, unlike the refined product which is highly volatile. The Biafra model turned the oil palm mills in various communities into refineries to refine petroleum products. They were able to produce diesel, kerosene, and gasoline and engine oil. The present administration could consider that model for local community refineries. Government should also deal with insidious and invidious cabal who do not want the government refineries to work because they are profiting from fuel importation and subsidy. Akin (2003) explained that the emergence of private refineries will create a better maintenance culture for the refineries and this will likely reduce unemployment by employing both skilled and unskilled labour. They would also engage in the training of manpower in Nigeria which will contribute to human capital development in the country. As for the urgent need for funds to develop the country's infrastructure it is suggested that government should source funds to finance capital projects elsewhere instead of relying on the proceeds from oil subsidy removal. One of such sources could be the Excess Crude Account Savings to finance Federal Government's capital projects

5.0 Conclusion

The knowledge that petroleum product prices in Nigeria are relatively low compared to prices in surrounding countries has also encouraged the government into reviewing the prices. The last reviews resulted in prices of key petroleum products increasing. The increases in prices are achieved by removing the subsidies on both

---

[11] If "Biafra" can refine her own petroleum products during the Nigerian Civil War (1967 – 1970) to meet her needs through communities' local refineries, Nigeria can complement her domestic fuel needs through local refineries instead of relying on importation.

Chapter Two | Ogujiuba Kanayo, in
Jideofor Adibe (Ed.)
The Politics and Economics of Removing Subsidies on Petroleum Products in Nigeria
London & Abuja, Adonis & Abbey Publishers

64

imported petroleum products as well as those produced in the country. By doing this, the twin problems of inefficiency in the sub sector and fiscal pressure are to be attended to. Subsidy removal will bring prices to an efficient level as well as make the sub sector more attractive to private local and foreign investors.

Evidence from around the world has shown that government owned enterprises are not as efficient as privately owned and controlled enterprises. This has informed the view that government had better restrict its role to that of providing an enabling environment for the private sector operators to function. These expectations , in as much as they are beneficial , are welcome as any policy to improve one or more sectors of the economy is needed as Nigerians have been experiencing declining average well-being i.e. poverty measures. These envisaged benefits are sectoral and macroeconomic in nature. However, the failure of the top-down approach has questioned the expected transmission of macroeconomic benefits to the household level and consequently their eventual effect on household poverty.

Full deregulation of the petroleum sector is something that should happen, but sequenced. Nigeria, with its endemic official and non-official corruption, inefficiency, profligacy, mismanagement and undemocratic system, the government can hardly be trusted to implement the gains of such removal and deregulation to the benefit of its over 160 million people. Can governments at all levels, captains of industry and businessmen be trusted to transmute the gains from subsidy removal into improving the welfare of the masses, providing basic infrastructure such as electricity, roads, healthcare, education, jobs, and general well-being expected of an oil-rich nation?. The government is not convincing either. Government is yet to tackle corruption in the oil sector and indeed in all the sectors of governance in the country to a convincing point, which would thus warrant a conviction on the action plans for the gains from removal of the subsidy and the palliative measures to ease the pains of subsidy

Chapter Two | Ogujiuba Kanayo, in
Jideofor Adibe (Ed.)
The Politics and Economics of Removing Subsidies on Petroleum Products in Nigeria
London & Abuja, Adonis & Abbey Publishers

removal. At the present, there is no guarantee that the savings to be realized from the oil subsidy removal will be used prudently and for the benefit of all Nigerians and not just a few corrupt and greedy individuals.

In spite of the possible positive macroeconomic effects of subsidy removal, there are unanswered questions. The implementation of a 'gradual deregulation' of the petroleum sub sector therefore brings up certain issues pertinent to the country's drive towards growth with poverty reduction: Can the country sufficiently monitor the impacts of this chosen pattern of efficiency through- deregulation? Can she follow paths that minimize possible adverse effects? With cautious optimism the preceding questions can be answered in the affirmative. Certain pertinent questions arise from the above issues:

a) Will the increases lead to higher poverty rates?
b) What role will these increases play in the dynamics of poverty in Nigeria?
c) Which socioeconomic groups and sectors will be most affected?

Unless the above issues are resolved, it is plausible that as gradual subsidy removal is achieved other unanticipated adverse socio-economic effects will follow[12].

By varying the composition of expenditure, government can carefully aim at using the increased total revenue arising from the subsidy removal to target expenditure compositions which can best counter negative effects in particular and poverty in general. Also, the relative effects of the petroleum subsidy removal on different socioeconomic groups should be explored before implementation, to

---

[12]A strategic economy wide view of the subsidy removal is needed, which support policies such as government expenditure level and composition will control to mitigate any adverse effects of the increases.

Chapter Two | Ogujiuba Kanayo, in
Jideofor Adibe (Ed.)
The Politics and Economics of Removing Subsidies on Petroleum Products in Nigeria
London & Abuja, Adonis & Abbey Publishers

avoid ambiguous responses. This will go a long way in carrying out the increases in ways that minimize or avoid the adverse effects. These questions have to be answered in order to design a price deregulation process that does not end up being well intended *yet* having a negative net-effect in real terms on the welfare of the majority of Nigerians. This can be done by anticipating both the positive and negative impacts so that the former are consolidated and the later countered with appropriate policy. Furthermore, it is recommended that incorporating information on the dynamics and heterogeneity of household income sources would be beneficial to the government in the sequencing of the proceeds of the subsidy removal.

References

Abel, B. and Bernanke, B. (1992): *Macroeconomic and Stabilization Policy* (Addison-Wesley Publishing Company).

Abiola, A. G. (2010): 'Analysis of the Impact of Removal of Subsidy on Petroleum Products on Public

Abubakar N.K (2011): 'The Return of Fuel Subsidy Removal Debate', Friday Business Day 15 July.

Adejumo Akintokunbo (2011): 'The Return of Fuel Subsidy Removal: Fraud, Deception, Corruption or Good-intentioned?', http://www.nigeriansinamerica.com/articles/5428/1/Fuel-Subsidy-Removal-Fraud-Deception-Corruption-or-Good-intentioned/Page1.html (Accessed: 21 March, 2013)

Adenikinju, Adeola (2000): 'Analysis of Energy Pricing Policy in Nigeria: An Application of a CGE Model', Research for Development, (NISER)

Askunowo V.O (2012): 'The Economics of Nigeria's Petroleum Products' Subsidy Removal Debate: Who is Right? Who is wrong'? *OPEC Energy Review*, Vol. 36 Issue 3.

Hossain, Shahabuddin (2003), 'Taxation and pricing of petroleum products in developing countries: A framework for Analysis with

Chapter Two	Ogujiuba Kanayo, in
	Jideofor Adibe (Ed.)
	The Politics and Economics of Removing Subsidies on Petroleum Products in Nigeria
	London & Abuja, Adonis & Abbey Publishers

Application to Nigeria, http://www.imf.org/external/pubs/ft/wp/2 003/wp0342.pdf (Accessed 21 March 2013)

IMF (2003): 'Issues and Prospects in the Oil and Gas Sector in Nigeria: Selected Issues and Statistical Appendix' Country Report No. 03/60.

Mukhtar B. (2012): 'Fuel Subsidy Removal Debate: The Way Forward', //www.gamji.com/article9000/NEWS9588.htm

Nigerian Bureau of Statistics, NBS, (2011): *Nigerian Poverty Profile Report* (Abuja, NBS)

Nwachukwu S.C. (2011): 'Which Petroleum Subsidy'? *Business Day* 28 June.

Nwadialo, U. (2012):'Fuel subsidy removal: A Nigerian Dilemma, *Vanguard*, and January 9.

Nwafor M; Ogujiuba K & Asogwa R (2006): 'Does Subsidy Removal Hurt the Poor'; *SISERA Working Paper Series*; 2007-16.

Ogunbodede, E.F Ilesanmi A.O and Olurankinse F. (2010): 'Petroleum Motor Spirit (PMS): Pricing Crisis and the Nigerian Public Passenger Transportation System', the *Social Sciences*, Volume 5, Issue 2, 113-121

Revenue in Nigeria', PARP Policy Analysis Report, (Abuja, PARP, and Abuja)

Chapter Two | Ogujiuba Kanayo, in
Jideofor Adibe (Ed.)
The Politics and Economics of Removing Subsidies on Petroleum Products in Nigeria
London & Abuja, Adonis & Abbey Publishers

68

CHAPTER THREE

A Case for Fuel De-Subsidization in Nigeria

Benedict Ndubisi Akanegbu

Introduction

This chapter examines the economic effects of the deregulation of the downstream oil sector, which to the average Nigerian, simply means the removal of fuel subsidy on petroleum products in Nigeria. The state of the Nigerian economy during the last fifty years has not been unlike that experienced in many developing countries. Agriculture prior to independence (1960) was the mainstay of the Nigerian economy, supplying the needed food requirements and employment opportunities for the majority of the labor force. It provided raw materials for the industrial sector and was also the chief foreign exchange earner for the country. Prior to the oil boom (1960-1973), the contribution of virtually all sectors of the economy grew rapidly. However, in terms of sectoral contribution to the gross domestic product (GDP) during this period, agriculture was the most important component of the economy. However, during the period (1974-Present), agriculture's share of the GDP declined drastically, and the share of the manufacturing sector in the GDP was still negligible, while that of oil increased tremendously. Presently, with over 36 billion crude oil reserves and average daily production of 2.6 million barrels of crude oil per day, Nigeria is the seventh biggest exporter of crude oil with 10th largest reserves in the world, but the only country among OPEC that imports refined petroleum products. During this period (1974-present), oil dominated the country's economic and financial performance to such a degree that changes in the fortunes of this single commodity affected significantly all sectors of the economy. Oil displaced agriculture as the major foreign

Chapter Three | Benedict N. Akanegbu, in
Jideofor Adibe (Ed.)
The Politics and Economics of Removing Subsidies on Petroleum Products in Nigeria
London & Abuja, Adonis & Abbey Publishers

exchange earner for the country and the prime mover of the economy. Therefore, the engine of Nigeria's economic growth during this period is its oil exports. These oil resources entailed a double-barrelled effect on the Nigerian economy. On one hand, they facilitated economic growth. On the other hand, they unleashed inflationary and distortionary tendencies. However, in order to slow down the impact of the resulting inflation, the government enforced a policy of price controls on a wide variety of goods especially on petroleum products. These policies caused distortions that have left the economy with several misallocated resources. This system of fuel subsidy on petroleum products entailed by the price controls provided wrong signals to both public and private enterprises, causing significant inefficiencies in the allocation of resources.

In the present context, economic growth will depend on the pace and effectiveness of the policy reforms, which should be designed to eliminate the distortions in the economy. Needless to say, reforms of the pricing policy should, therefore, constitute a major component of any remedial program. The experience over the past few years has shown that the removal of fuel subsidy in macroeconomic and sectoral policies is an essential step in controlling the deteriorating state of the Nigerian economy. However, the removal of fuel subsidy is a necessary but not a sufficient condition for achieving sustainable economic growth. It is necessary since it will help generate high growth rates in the external sector, but may not be sufficient if such high growth rates occur only in one sector of the economy that is not linked to the others.

Background of the Study

Prices in general are crucial to the operation of any economic system, especially a private market economy. Price policies are judged by their effects on promoting economic efficiency and hence faster growth of income. Each price policy uses a subsidy or a tax or a trade restriction to cause the domestic producer or consumer price, or both,

Chapter Three | Benedict N. Akanegbu, in
Jideofor Adibe (Ed.)
The Politics and Economics of Removing Subsidies on Petroleum Products in Nigeria
London & Abuja, Adonis & Abbey Publishers

to differ from the world price. For instance, tariffs discriminate against foreign producers in favor of domestic producers. The same kind of discriminatory treatment is accorded through subsidies to domestic producers though with quite different other effects. Subsidies may therefore be regarded as a substitute for import tariffs as a means of favoring domestic industries over foreign competitors. Subsidies may be used not only to reduce imports but also as a method of increasing exports. Export subsidies permit affected commodities to be offered on the world market at a lower price than domestically, thus encouraging a larger volume of exports. A price policy intervention affects economic growth by the extent of efficiency losses. However, a simple consumer subsidy causes both producers and consumers to face lower prices than those in the world market. A specified producer price subsidy can raise the decision price above world levels while leaving the consumer price at the world price. A consumer subsidy on importable is a common price policy intervention in developing countries. For instance, subsidizing fuel imports causes the domestic price of fuel for both consumers and producers to be less than the world price. As a result, the quantity of fuel produced domestically is below what it would have been without the subsidy.

Chapter Three | Benedict N. Akanegbu, in
Jideofor Adibe (Ed.)
The Politics and Economics of Removing Subsidies on Petroleum Products in Nigeria
London & Abuja, Adonis & Abbey Publishers

Price

Domestic demand   Domestic supply

Ps ·· Self-sufficient price

Pw A   B   C   D   E ············ World price

Pd ·· Domestic price
      G   H

O   Q2 Q1   Q3 Q4   Quantit

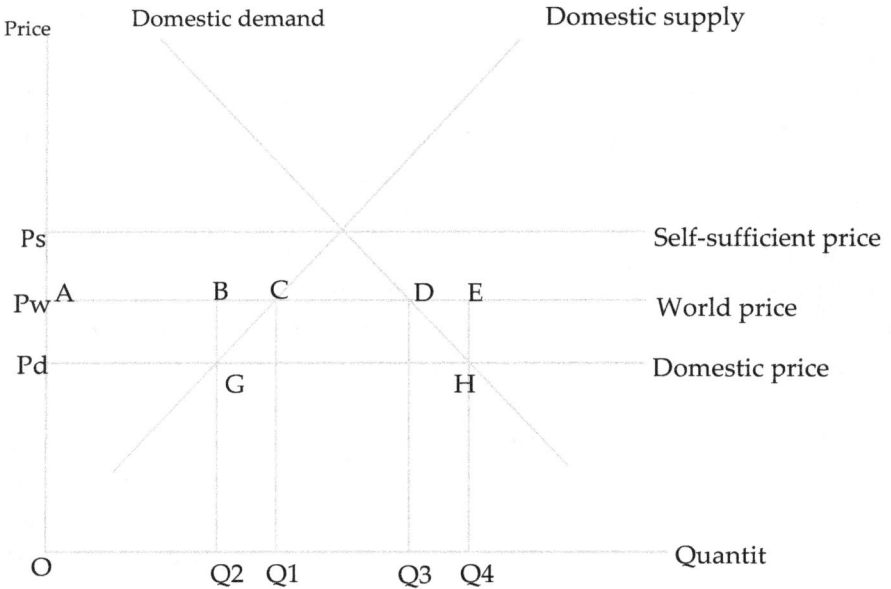

Figure 1. Effects of Fuel Subsidy Policy

From figure 1 above, it can be seen that before the subsidy policy on fuel, the domestic price of fuel is equal to the world price of fuel, and so the domestic supply of fuel is Q1 and domestic demand of fuel is Q3, with imports making up the difference (Q3-Q1). But when the government introduces a fuel subsidy which lowers the domestic price of fuel below the world price of fuel by Pw-Pd, the demand of fuel product increases to Q4 and the domestic supply of fuel drops to Q2, both because of lower fuel prices; and the import gap widens to Q4 – Q2. In the above analysis, there are efficiency losses in both the producing and consuming countries caused by the wedge between domestic and international fuel prices. The production efficiency losses are measured by the triangle BCG, while the consumption losses are measured by the triangle DEH.

Chapter Three | Benedict N. Akanegbu, in
Jideofor Adibe (Ed.)
The Politics and Economics of Removing Subsidies on Petroleum Products in Nigeria
London & Abuja, Adonis & Abbey Publishers

In both developed and developing countries, fuel subsidies exist in different variety of ways such as direct budgetary transfers, tax exemptions and rebates. Price controls, market access limits and trade restrictions are also key elements of fuel subsidies (Varangu and Morgan, 2002). Fuel subsidies provided through direct financial transfers (including tax rebates) are sometimes referred to as direct transfers, while those provided through other mechanisms are often referred to as indirect transfers. Fuel subsidies alter fuel prices, causing market distortions. These distortions have wide economic implications by distorting prices and therefore affecting production and consumption decisions, and in most cases straining government budgets by imposing fiscal burdens, which in turn reduces the amount of money available to spend on social programs (Saunders and Schneider, 2000). Also, it has the capability of diverting funds that could have been spent on healthcare or education (Koplow and Dernbach, 2001).

In theory, removing fuel subsidies would result in higher fuel prices in countries that currently subsidize consumer prices, which would reduce consumption and also remove a costly drain on the government budget. Fuel subsidies come in two main forms: those designed to reduce the cost of consuming fuels; and those aimed at supporting domestic fuel production (Burniaux et al., 2009). Subsidies aimed at consumers are generally intended to keep fuel prices low to alleviate poverty by expanding the population's access to energy (Morgan, 2007). These types of subsidies are more common in developing countries. And it takes the form of price controls. Subsidies aimed at producers generally keep costs of production lower or increase revenues, and their effects is to keep marginal producers in business (Saunders and Schneider, 2000). Economic theory states that social welfare is maximized when the price of each good and service is freely determined by the interaction of buyers and sellers in open, competitive market. Therefore, subsidies can be justified if overall social welfare is increased. However, experience

Chapter Three | Benedict N. Akanegbu, in
Jideofor Adibe (Ed.)
The Politics and Economics of Removing Subsidies on Petroleum Products in Nigeria
London & Abuja, Adonis & Abbey Publishers

has shown that net effects of subsidies are negative. And this means that the overall social welfare would be higher without subsidies (Ellis, 2010). Hence, the argument in favor of fuel subsidy removal becomes very strong.

A review of official documentation and media reports reveals that the issue pertaining to the removal of fuel subsidy in Nigeria has been a constant debate by experts, politicians and ordinary Nigerians over a period of time. Past administrations have been involved in one form of petroleum products price increment or the other. For instance, there have been 11 adjustments of the pump price of fuel between 1998 and 2009. While the late General Sani Abacha pegged the pump price at N11 per litre throughout his five-year reign from 1993 to 1998, the price was adjusted to N19 during the administration of General Abdulsalam Abubakar, which lasted for nine months. Former President Olusegun Obasanjo adjusted the price eight times during his eight-year reign between 1999 and 2007, from N19 per litre to N70 per litre, before the late President Musa Yar'Adua adjusted it downwards to N65 per litre. As at August 15, 2011, based on its pricing template, Petroleum Products Pricing Regulatory Agency (PPPRA) said that the landing cost of a litre of petrol in Nigeria is N129.2; the margin for transporters and marketers is N15.49; the export pump price therefore ought to be N144.70 per litre instead of the N65 per litre that is charged. This means that there is a subsidy of N79.70 assumed by government on every litre of petrol sold in Nigeria. The deregulation policy in form of the Petroleum Industry Bill (PIB) before the National Assembly is based on the report of the oil and Gas reform implementation Committee (OGIC) set by the Federal Government in 2000 to carry out a comprehensive reform of the oil industry. It provides the new legal framework for the organization and operation of the entire oil industry in Nigeria. The PIB recommends the deregulation of the industry to allow market forces to determine the prices of petroleum products. It is also true that as the price of oil goes up at the international market, the price of

Chapter Three | Benedict N. Akanegbu, in
Jideofor Adibe (Ed.)
The Politics and Economics of Removing Subsidies on Petroleum Products in Nigeria
London & Abuja, Adonis & Abbey Publishers

petroleum also goes up in the country, resulting in constant adjustment of the pump price of fuel.

The Economic Effects of Fuel De-subsidization:

The economic effects of the removal of fuel Subsidy are numerous. However, we shall concentrate our analysis on fuel subsidy removal as it affects Nigerians and the Nigerian economy in particular.

Price-cost effects

The immediate impact of petroleum subsidy removal is on the price and costs of petroleum. There are other, more fundamental consequences, but they are brought about, in a market price system, through the mechanism of changing price and cost relationships. The most general price effect of petroleum subsidy removal is to correct the differential in the international price (and cost) of petroleum products. Without subsidies, assuming pure competition and ignoring transportation costs, we know that the price (and cost) of petroleum will tend to be the same with the trading countries. But with the introduction of petroleum subsidy, the price of petroleum becomes cheaper than in the other trading countries. And the price differential is equal to the amount of the subsidies in most cases.

The reason that a subsidy creates an international price and cost differential equal in amount to the amount of the subsidy can be easily shown as follows: Suppose that the Nigerian government is subsidizing imported petroleum products from Ghana by 50 naira per liter. Then, under competitive conditions, the equilibrium price of imported petroleum is 50 Naira lower in Nigeria than in Ghana. For if the price in Nigeria, after the imposition of subsidy is 50 Naira lower than in Ghana, importers would lose on petroleum imports. Hence, the import of petroleum would be reduced, and its price in Nigeria would rise. On the other hand, if the petroleum commanded a price in Nigeria higher than in Ghana by more than the 50 Naira,

Chapter Three | Benedict N. Akanegbu, in
Jideofor Adibe (Ed.)
The Politics and Economics of Removing Subsidies on Petroleum Products in Nigeria
London & Abuja, Adonis & Abbey Publishers

the profits realized by importers would stimulate greater imports until the price falls.

Suppose that a country produces at home some part of its domestic consumption of a commodity and imports the remainder from another country. This implies that the commodity is produced under increasing-cost conditions, with the supply curves in both countries rising from left to right on the conventional diagram. In competitive equilibrium and ignoring transportation costs, the price (and costs) of the commodity are the same in both countries. Now let the importing country impose a subsidy on the commodity. Several reactions will follow: First, as an immediate reaction, the price of the commodity in the importing country will tend to fall, and the quantity demanded to rise. Second, domestic producers of the good in the importing country will be induced by the lower market price to reduce their output. Third, producers in the exporting country will find that the export market has expanded – both because of the smaller domestic output in the importing country and because of the greater quantity demanded in the importing country. Therefore, the output in the exporting country will tend to rise. Fourth, the costs (average and marginal) of production will fall in the importing country with its decreased output and rise in the exporting country with its increased output. Hence, in the new equilibrium position, the price of the commodity will be lower in the importing country and higher in the exporting country.

The above analysis will help to explain the petroleum subsidy situation in Nigeria: Nigeria produces some part of its domestic consumption of petroleum products and imports the remainder from other countries. This means that petroleum product in Nigeria is produced under increasing-cost conditions. The difference between the export price and the domestic price is paid by the government to the few petroleum importers as a windfall gain to those fortunate enough to have received import licenses from the government. Therefore, with the removal of fuel subsidy; there will be an

Chapter Three | Benedict N. Akanegbu, in
Jideofor Adibe (Ed.)
The Politics and Economics of Removing Subsidies on Petroleum Products in Nigeria
London & Abuja, Adonis & Abbey Publishers

immediate hike in the pump prices of fuel and transport fares in the short run. And this may generate an inflationary spiral in the economy through both a cost push and demand pull inflation. However, in the long run; the market forces of demand and supply will correct all the misallocation of resources. And the fuel price will be reduced to reflect the competitive market price.

The effects on the volume and benefits of trade

From our preceding analysis it is clear that one of the major effects of subsidy is to reduce the volume of international trade. This effect follows from the fact that, the price of a commodity subject to subsidy is reduced in the importing country, thereby reducing the total quantity produced domestically in the short run. And the reduction in the volume of import is directly brought about by the quantitative limits on the volume of imports permitted. Since international trade yields benefits to the world in the form of higher real income, subsidies, by reducing the volume of trade, presumably lower the world's real income. The loss involved derives not simply from the quantitative reduction in trade but also from the misallocation of resources. There is overwhelming empirical evidence that suggests a strong link between price distortions and economic growth especially in developing countries. Harberger (1959) attempted to explore the possible results of eliminating misallocations of resources in economies like Chile, Brazil, and Argentina. It was concluded that policies aimed at eliminating distortions in the price mechanism can raise the long-term rate of growth of national income. However, it is believed that as resources are deployed more productively, there will be terms of trade effects. Terms of trade is the ratio of the price a country receives for its exports to the price it pays for its imports expressed as a percentage. An improvement in a country's terms of trade is generally understood to improve a country's social welfare. In fuel importing countries such as Nigeria, removing fuel subsidies implies a term of trade gain that comes in addition to the gains from

Chapter Three | Benedict N. Akanegbu, in
Jideofor Adibe (Ed.)
The Politics and Economics of Removing Subsidies on Petroleum Products in Nigeria
London & Abuja, Adonis & Abbey Publishers

77

the subsidy reform. Oil producing countries record term of trade losses from removing subsidies (Burniaux et al., 1992). At this point it will be advisable to note that in as much as Nigeria is one of the oil exporting countries, it still remains the only one that imports refined oil. Therefore, the removal of fuel subsidy from petroleum products will bring about increased volume of trade in the Nigerian economy and will eventually have a beneficial effect on Nigerians.

The effects on the allocation of resources

Effective exchange depreciation also reduces the volume of imports, but it is a reduction that is compelled by the requirements of equilibrium. In contrast to subsidies, exchange rate depreciation does not distort cost-price relationships but rather brings them into new alignment. By generally rendering imports more expensive to the depreciating country and by causing the prices of its exports to become cheaper to foreign buyers, exchange depreciation tends to contract the volume of imports and to increase the volume of exports. The result of free and unrestricted international trade is to eliminate (except for transport costs) international differences in the price and cost of traded goods and services. But now we have seen that subsidies artificially reimpose international price-cost differences. This international comparative cost differences create the opportunity for beneficial international trade; subsidies "freeze" this opportunity, not allowing it fully to be taken advantage of. The extensive employment of subsidy results, then, in a virtual breakdown of the price mechanism as a guide in the international allocation of resources. Subsidies have to be evaluated strictly by economic principles of resources allocation. Bhagwati (1978) and Timmer (1980) have stressed the existence of potentially high social costs of domestic price distortions in terms of their resources allocation, national output, and income distribution effects. And Aguirre and Yucelik (1981) in their review of African experience emphasized that mixing of revenue and protective functions has led to excessive levels

Chapter Three | Benedict N. Akanegbu, in
Jideofor Adibe (Ed.)
The Politics and Economics of Removing Subsidies on Petroleum Products in Nigeria
London & Abuja, Adonis & Abbey Publishers

of protection resulting in damaging effects on resources allocation. In any case, what is crucial in this respect is the need for empirical cost-benefit studies in order to determine whether resources are being fairly and economically allocated. Although economic nationalism is an important strategy towards the minimization of chronic dependency, one cannot pursue economic nationalism to the point of total disregard for the principle of comparative advantage in international trade or domestic factor mobility. Subsidization in this category of activities tends to reinforce inefficiency in production, monopolistic control, poor and uncreative management as well as the diminution of the competitive drive. Failure to find an optimal balance between these operative forces would produce chronic distortion in the allocation of scarce resources to alternative uses (Anise, 1980). Results from a wide variety of global and single country economic modeling studies of subsidy reform suggest that on an aggregate level, changes to GDP are likely to be positive due to the incentives resulting from price changes leading to more efficient resource allocation (Von Moltke et al., 2004). Subsidies distort prices, and fail to reflect the true costs of supply and therefore affect resource allocation decisions, production and consumption (Saunders and Schneider, 2000). Removing fuel subsidies would enhance Nigeria's growth prospects through greater allocative efficiency and fiscal benefits, although there would be significant adjustment costs in the short term.

The removal of fuel subsidies in the Nigerian economy will have a positive effect on efficient resource allocation decisions affecting production and consumption decisions of Nigerians. This will have a beneficial effect on international trade with positive changes on the gross domestic product (GDP) of Nigeria.

Chapter Three | Benedict N. Akanegbu, in
Jideofor Adibe (Ed.)
The Politics and Economics of Removing Subsidies on Petroleum Products in Nigeria
London & Abuja, Adonis & Abbey Publishers

79

Consumption effects

The effects of subsidies are not confined to distorting the allocation of resources and the accompanying loss of some of the specialization benefits of trade. Consumption is also distorted, and the full exchange benefits of trade are prevented from being realized. Just as the loss from inefficient allocation as a result of trade restriction is the counterpart of the gain from specialization under free trade, so the loss to consumers as a result of trade restrictions is the counterpart of the exchange benefits of free trade. Generally, the creation of artificial price differences through subsidy forces consumption into patterns which yield less total utility than is yielded by given expenditure when prices are equalized internationally.

Subsidies can increase fuel consumption and reduce incentives for fuel efficiency. Subsidies that reduce prices for consumers promote higher consumption of fuel, and reduce incentives to use fuel efficiently (Morgan, 2007). Subsidies that lower end-use prices, either directly or by lowering the cost of production, always lead to higher fuel use (except when supply is rationed), and reduce incentives to conserve or use fuel more efficiently. The extent of the increase in consumption depends on the price elasticity of demand. Overall, in the short-run, removing fuel subsidies will increase price levels and reduce household consumption. Both high and low income groups are affected. By contributing to macroeconomic stability, subsidy removal should over the long-run be beneficial to the poor. Therefore, the removal of fuel subsidies will encourage average Nigerians on the efficient use of petroleum products thereby enjoying the opportunity of maximizing their individual utilities in fuel consumption.

Chapter Three | Benedict N. Akanegbu, in
Jideofor Adibe (Ed.)
The Politics and Economics of Removing Subsidies on Petroleum Products in Nigeria
London & Abuja, Adonis & Abbey Publishers

Effects on the distribution of income

It will be recalled that one of the effects of international trade, conducted in the context of the Heckscher-Ohlin model, is a redistribution of income in favor of the owners of the relative abundant factor of production and away from the owners of the relative scarce factor of production. Subsidies reverse the direction of this redistribution, favoring the relatively scarce factor at the expense of the relatively abundant factor. In addition to the above general effects of subsidies on income distribution, particular factors of production that are specialized and not readily transferable among different employments are prone to be harmed, or benefited, by subsidies. For example, Nigeria has a comparative disadvantage in the labor-intensive petroleum industry and that petroleum workers have specialized skills and training not useful in other industries. The introduction of free trading relations with other countries would tend to cause a contraction in domestic production of petroleum and the discharge of petroleum workers, who will either remain unemployed or be forced into lower-wage occupations. Subsidization on the import of petroleum on the other hand, would preserve the market for the domestic industry and therewith the jobs and wages of petroleum workers. To the extent that reduced imports lead to reduced exports, subsidization of petroleum products cause employment opportunities and wage rates in export industries to contract. Higher fuel prices would raise the cost of living and producer costs. The cost of living would rise directly and indirectly through the effect of increased fuel-input costs on the price of consumer goods. Therefore, the removal of fuel subsidies in Nigeria will create more employment and likely increases in wage rates in export industries.

The present and continuing problems are first, how to achieve equilibrium between increasing government expenditures and widely fluctuating government revenues, and second, how to achieve equitable redistribution of income so as to reduce apparent social

Chapter Three | Benedict N. Akanegbu, in
Jideofor Adibe (Ed.)
The Politics and Economics of Removing Subsidies on Petroleum Products in Nigeria
London & Abuja, Adonis & Abbey Publishers

polarization, while taking effective measures against inflation and striving towards self-sufficiency in food and agricultural production. Their solution therefore involves the dynamic interplay of group interests and power and influence relationship in the society (Anise, 1980). There exists "the parasitic group" (the so called cabals) that makes easy money in a narrowly circumscribed enclave of fuel subsidy, that confers great boon on a few who are strategically poised in the web of governmental power and influence. This culture legitimizes increasingly polarizing structures of resources allocation (Anise, 1979).

A policy of income redistribution can mean redistribution downwards or redistribution upwards. In Nigeria, income redistribution has tended to be upwardly flexible but downwardly and horizontally rigid. In effect, the top rich and powerful tend to get richer and more powerful. The poor lower echelon wage earners tend to stagnate while the middle is squeezed into chronic anxieties and frustrated dreams of the good life. Thus, the policy cannot ultimately solve the problem of growing income inequalities in Nigeria. If the problem is not solved, then the social polarization will tend to rise rather than decline. The rise of socioeconomic polarization will tend to worsen the problems associated with the rapidly rising class structure, vested class interest, and emerging class struggle in the society. Therefore, making a case for a radical public policy directed towards achievement of equitable income distribution may be permissible only within the purview of this persistent illusion of social justice as a fundamental purpose of governance in any society. Hence, a radical policy of de-subsidization is a most effective way to achieve the stated policy objectives within a generation (Anise, 1980).

Effects on the balance of payments

Apart from the effects of subsidies, and related policies on the allocation of resources and on consumption, they may also be a source of balance-of-payments disequilibrium and an impediment to

Chapter Three | Benedict N. Akanegbu, in
Jideofor Adibe (Ed.)
The Politics and Economics of Removing Subsidies on Petroleum Products in Nigeria
London & Abuja, Adonis & Abbey Publishers

the adjustment mechanism. Starting from a position of equilibrium, an increase in subsidies by one major country or group of countries reduces the exports of other countries, causing a balance-of-payments deficit and necessitating readjustments of a structural character.

Subsidies always have an impact on international trade. Consumption that increases energy use boosts demand for imports or reduces the amount of fuel available for export. This harms the balance of payments by increasing the country's dependence on imports. For example, the massive increase in fuel use in Nigeria that resulted largely from subsidies has led to a huge increase in imports of refined petroleum due to excessive demand. Removing fuel subsidies would improve the balancing of the national budget substantially. In addition, removing subsidies would provide Nigeria with a hedge against exchange-rate fluctuations. And will have a positive effect on the balance of payments through its positive effects on exports since the amount of fuel available for exports will be increased. The removal of subsidies to fuel products would lower domestic demand in Nigeria and free more fuel for export which in turn will improve the balance of payments position of Nigeria.

Conclusion

Nigeria has over the years been engaged in a wide range of price interventions in the petroleum sector with the sole intention of providing incentives to promote sectoral growth. However, these pricing and subsidy policies have had a distorting impact on the price-cost relations, volume and benefits of trade, allocation of resources, consumption, distribution of income, and balance of payments within the sectors, thereby generating considerable costs in terms of economic efficiency. Since future economic growth depends on the pace and effectiveness of policy reforms designed to eliminate the price distortions in the economy, reforms of pricing policy (Fuel Subsidy removal) should constitute a major component of any remedial program. The analysis in this paper strongly supports the

Chapter Three | Benedict N. Akanegbu, in
Jideofor Adibe (Ed.)
The Politics and Economics of Removing Subsidies on Petroleum Products in Nigeria
London & Abuja, Adonis & Abbey Publishers

conclusion that there are significant economic benefits that would result from the removal of fuel subsidy on petroleum products in Nigeria. On this basis, there is a mounting body of evidence that policy-makers should not wait to begin the implementation process of fuel subsidy removal. The policy of gradualism cannot succeed. What is needed is a bold initiative, a true confrontation with destiny by immediate de-subsidization of fuel prices. The issue is that the combination of fiscal and monetary policy measures now being adopted will at best contain these problems but not solve them. However, de-subsidization will make full economic and social-political sense only if it is carried out within the structure of a well-designed income policy predicated on the maximization of social equity and efficiency in resource allocation (Anise, 1980). According to the federal government of Nigeria; it expects to free up to N1.2 trillion in savings from the expected removal of fuel subsidies, and the efficient re-allocation of these freed resources particularly to the transportation and social sectors may mitigate the harsh effects of the policy in the short run. It is also argued that if fuel subsidy is removed, the petroleum sector will be opened up for more competition which will lead to building more refineries and the law of demand and supply would come into play, and at the long run, prices of fuel would fall.

The only argument against the subsidy removal is that deregulation can only lead to even higher prices and inflation of virtually every commodity which would cause more hardship for the ordinary Nigerians who are most likely to be hit by the inevitable hike in the prices of petroleum products and the increased costs of living that may be induced by the decision in the short run. And this skyrocketing of the prices of petroleum products is sure to induce anger across the land. Also, majority of Nigerians tend to believe that the fuel crisis is a war stratagem of artificial scarcity deliberately created by the government in order to justify price hikes and the removal of fuel subsidies. And what the government does with its

Chapter Three | Benedict N. Akanegbu, in
Jideofor Adibe (Ed.)
The Politics and Economics of Removing Subsidies on Petroleum Products in Nigeria
London & Abuja, Adonis & Abbey Publishers

gains from Nigerian's oil earnings is still very questionable (Aina and Odebiyi, 1998).

Policy Recommendations

Fuel De-subsidization can be argued to be the way forward in expanding opportunities for economic growth and competitive markets. Hence, the issue is not whether the de-subsidization of fuel products is a good economic policy or not but the main issue at hand is the timing of the policy implementation. Therefore, it should be more appropriate and sensible to delay the removal of the fuel subsidy until the federal government delivers on a number of strategic measures to alleviate the impact on the average Nigerians as follows:

- Put pressure on the government to resuscitate the four existing refineries to work to full operational capacity given the implications for overhead and competitiveness for local industries.
- Build new refineries so that the idea of importing refined products will stop and at the same time be able to refine more of the crude so as to give the country higher stakes in a deregulated market, stimulate medium scale service industries and also provide greater job opportunities for the teaming skilled unemployed youths.
- Put in place a mechanism of monitoring the quantity of crude oil produced per day in the country and stop illegal bunkering and theft of oil in the Niger Delta to reduce the loss of revenue and use the savings to fix old refineries and build new ones.
- Restructure and fight corruption in oil industry and at NNPC in line with its counterparts – Petronas in Malaysia and Petrobas in Brazil so as to make the products available to Nigerians at a cheaper rate.

Chapter Three | Benedict N. Akanegbu, in
Jideofor Adibe (Ed.)
The Politics and Economics of Removing Subsidies on Petroleum Products in Nigeria
London & Abuja, Adonis & Abbey Publishers

- Cut down on government spending - the huge wastages at the various levels of the federal bureaucracy (the amount spent on public office holders).
- Encourage the rehabilitation and building of the following infrastructures: roads, schools, hospitals, portable water, and electricity supply required to service industries.
- Introduction and execution of social programmes, rehabilitating the refineries, and job creation through the savings realized from the expected removal of fuel subsidies by creating a special fund solely for that purpose.
- Finally, government should push for greater accountability and good governance to ensure for a more transparent privatization process that will respond to free market mechanism of demand and supply which will in turn have a 'trickle down' effect on the economy.

References

Aguirre, P.S. and Yucelike, M. (1981): *Tax Policy and Administration in Sub-Saharan Africa,* Washington: International Monetary Fund.

Anise, L. (1979): 'Confrontation Politics and Crises management: Nigerian University Students and Public Policy'. *Issue: A Journal of Opinion,* 9(1): 73-82.

Anise, L. (1980): 'De-subsidization: An Alternative Approach to Government Cost Containment and Income redistribution Policy in Nigeria', *African Studies Review,* vol. 23, No 2.

Aina, O.I. and Odebiyi, A.I. (1998): 'Domestic Energy Crises in Nigeria: Impact on Women and family welfare', *African Economic History* No. 26, PP. 1-14.

Bhagwati, J. (1978): *Anatomy and Consequences of trade control regimes,* Washington: national Bureau for Economic Research.

Burniaux, J.M., Chanteau, J., Dellink, R., Duval, R. and Jamet, S., (2009): 'the economics of Climate Change Mitigation: How to

Chapter Three | Benedict N. Akanegbu, in
Jideofor Adibe (Ed.)
The Politics and Economics of Removing Subsidies on Petroleum Products in Nigeria
London & Abuja, Adonis & Abbey Publishers

86

build the necessary global action in a cost-effective manner,' Economics Department Working Papers No. 701.

Burniaux, J.M., Martin, J.P. and Oliveira-Martins, J. (1992): 'the effects of existing distortions in energy markets on the costs of policies to reduce CO2 emissions: evidence from Green', *OECD Economics studies* 19 (Winter), 141-165.

Ellis, J. (2010): 'The Effects of Fossil-Fuel Subsidy Reform: A review of modeling and empirical studies', series of papers produced by the Global Subsidies Initiative (GSI) of the International Institute for Sustainable Development (IISD).

Harberger, A.C. (1959): 'The Fundamentals of Economic Progress in Underdeveloped Countries: Using the resources at hand more efficiently', *American Economic Review*, 49, 134-146.

Koplow, D. and Dernbach, J. (2001): 'federal Fossil-fuel Subsidies and Greenhouse Gas Emissions: A case study of Increasing Transparency for Fiscal Policy', *Annual Review of Energy and Environment*.

Morgan, T. (2007), 'Energy Subsidies: Their Magnitude, How they Affect Energy Investment and Greenhouse Gas Emissions, and Prospect for reform', Menecon Consulting.

Saunders, M. and Schneider, K. (2000): 'Removing energy subsidies in developing and transition economies', *ABARE Conference Paper*, *23rd Annual IAEE International Conference, International Association of Energy Economics*, June 7-10, Sydney.

Snider, D. (1979): *Introduction to international economics*, chap. 9, (Richard D. Irwin, Inc.,).

Timmer, C.P. (1980): 'Food Prices and Food Policy Analysis in LDCs', *Food Policy*, 5, 188-199.

Von Moltke, A., Mckee, C. and Morgan, T. (2004): *Energy subsidies: Lessons Learned in Assessing their Impact and Designing Policy Reforms*, (Sheffield: Greenleaf Publishing).

Chapter Three | Benedict N. Akanegbu, in
Jideofor Adibe (Ed.)
The Politics and Economics of Removing Subsidies on Petroleum Products in Nigeria
London & Abuja, Adonis & Abbey Publishers

CHAPTER FOUR

Subsidy as an Imperative for Sustainability in a Depressed Economy: A Case Study of Nigeria

Robert Madu and Shedrack Moguluwa

Abstract

Subsidy removal remains one of the issues that have generated heated debate among Nigerians irrespective of class, religion, political affiliations, orientation and academic status. The emotions and quick reactions to such issue are understandable as a very high percentage of Nigerian economy is dependent on petroleum. Earlier, it was coal and its dividends were used by the colonial masters to set up Nigerian administrative institutions together with white collar jobs, develop and build Nigerian cities. Afterwards, petroleum was discovered in great quantity within the Niger Delta region. Nigerian Petroleum became attractive in international market and yielding resources that the country has never witnessed before. This nonetheless became the platform upon which past and present governments of Nigeria hinged their developmental plans. Although the Nigerian government decided to maximize international sales/profit, they equally gave the citizens some leverage by way of subsidy. The euphoria of petroleum exportation with its attendant huge revenue streams overwhelmed the government which engaged extravagantly in white elephant projects. Instead of investing the oil yields in developmental projects, squandering and corruption took the center stage. The nations' economy shifted from agriculture which hitherto was the mainstay to petroleum products. The costs of major goods rose such that inflation rate doubled within few years. The government which at onset introduced petroleum subsidy to alleviate the suffering of the masses gradually started removing it as

Chapter Four | Robert Madu & Shedrack Moguluwa, in
Jideofor Adibe (Ed.)
The Politics and Economics of Removing Subsidies on Petroleum Products in Nigeria
London & Abuja, Adonis & Abbey Publishers

one of the strategies to salvage the ailing economy. This nonetheless negates the import and imperative of petroleum subsidy for sustainability in a depressed economy.

Key Words: *Subsidy, Sustainability, Depressed economy.*

Background

It was with mixed feelings that Nigerians received the news of the withdrawal of fuel subsidy on Premium Motor Spirit (PMS) by the President, Goodluck Jonathan in his official broadcast to the nation on the January 1, 2012; hiking the pump price of petrol to N141.00 ($0.94) per litre. Prior to this announcement, the pump price of PMS was N65.00 ($0.43) per liter as a result of the subsidy programme which the Federal Government of Nigeria established. The subsidy allows the government to bear part of the cost of the petroleum product so as to alleviate the burden of the price of consuming the product on Nigerian masses.

The removal of the fuel subsidy sparked off wide resentment across the country and culminated in a week-long industrial action with the organized labour and the generality of the Nigerian masses. A compromise between the Federal Government and the organized labour brought down the pump price of the PMS officially to N97:00 leaving the petroleum subsidy partially removed.

There is indeed a growing understanding among the elite and experts on the need to remove fuel subsidy in order to maximize the advantages of fuel price to provide better social services, but such understanding goes with characteristics which show that the timing can only be futuristic as a result of current low level of sustainability and the depressed nature of the Nigerian economy.

A crucial question about de-subsidization in a depressed economy, as Abang, Elufisan and Okwubunne (2012:126) put it was: "How has it all affected the common-man, (the middle income earner)

Chapter Four | Robert Madu & Shedrack Moguluwa, in
Jideofor Adibe (Ed.)
The Politics and Economics of Removing Subsidies on Petroleum Products in Nigeria
London & Abuja, Adonis & Abbey Publishers

who has been identified as the group of people in the nation, whose activities mostly drive economic growth and development?"

The Nigerian Economy (Subsidy and Depressed Economy in Perspective)

Subsidy, according to the Oxford Advanced Learners Dictionary connotes money that is paid by the government or an organization to reduce the cost of services or of producing goods so that their prices can be kept low. Depression on its own side, according to the same source means a period when there is little economic activity and many people are poor or without jobs, hence a depressed economy is deduced as an economy that lacks robust economic activities and many of its citizens are in poverty and unemployed.

Prior to the existence of the Nigerian nation as an independent entity in October 1960, her economy had been agro-dependent; surviving and thriving on the produce of her rich and fertile soil. Cash crops and food crops like oil palm, cocoa, rubber, cashew, groundnut, cotton, cassava, etc., have been the means of sustenance in export and domestic use for the Nigerian nation until the discovery of petroleum in commercial quantity in the Niger Delta region of Nigeria.

The agro-economy Nigeria had before the oil era gave her the opportunity to effectively plan and manage available resources. For one, food was sufficient and a number of infrastructural developments were financed through proceeds entirely earned from agricultural products, (Aliede, 2010).

The gift of nature that is meant to bless the land turned a curse as the oil boom of the early 1970s heralded gross mismanagement, ineptitude, recklessness and massive corruption. The extent of mismanagement resulted in an economy that totally neglected the usual agro proceeds and all eyes turned to the 'already-made money' flowing from the 'Black Gold' in the heart of the Niger Delta region. Before long, the implication of the negligent steps on agriculture

Chapter Four | Robert Madu & Shedrack Moguluwa, in
Jideofor Adibe (Ed.)
The Politics and Economics of Removing Subsidies on Petroleum Products in Nigeria
London & Abuja, Adonis & Abbey Publishers

91

began to dawn on the country that she started importing food from countries that she used to be of aid to.

Import bill for food in Nigeria is exceptionally high and it is growing at an unsustainable rate of 11% per annum. Ironically, Nigeria is importing what it can produce in abundance. This trend is hurting Nigerian farmers and displacing local production, (Partner, 2013).

The total dependence on petroleum gave rise to an economy that is controlled and determined by the price of petroleum products internationally and locally. Its direct consequence is the fact that the totality of consumption of every good and service in Nigeria is controlled by the pump price of petrol as the basic source of energy in the country.

The price of food, shelter, labour, transportation, communication and healthcare, being tied to the pump price of fuel helped to inspire the subsidization of the price of the product by the Federal Government of Nigeria to lessen the burden on the shoulders of the citizens. As Abang, et al, (2012:126) noted:

> The introduction of subsidy indirectly promotes economic growth and development as a result of the affordability of the price of goods which provides an enabling point for the middle class citizen to contribute significantly to the economy.

The crux of this piece therefore is that the fuel subsidy should (at this period) be sustained in Nigeria for the following peculiarities of the polity:

Corruption

> The government's own anti-corruption watchdog the Economic and Financial Crimes Commission (EFCC) estimates that between independence in 1960 and 1999, the country's rulers stole $400 billion in oil revenues – equal to all the foreign aid to Africa during the same period (The Nation, 2007).

Chapter Four | Robert Madu & Shedrack Moguluwa, in
Jideofor Adibe (Ed.)
The Politics and Economics of Removing Subsidies on Petroleum Products in Nigeria
London & Abuja, Adonis & Abbey Publishers

92

There are two broad types of corruption in this regard. Firstly, is the corruption in the petroleum industry which is evidenced in the report of the Okigbo Panel. The Dr. Pius Okigbo led Panel on the reorganization and reforms of the Central Bank of Nigeria reported that $12bn petroleum funds were mismanaged in 1994 (Okigbo, 2013).

And secondly, is the overwhelming social corruption in Nigeria; which the petroleum sector is just reflective of. Subsidy may have aided corruption, but its removal will not lead to the eradication of corruption. It will rather lead to another kind of corruption as past experiences have shown, like artificial scarcity of petroleum products and hoarding of fuel. What is required is a change of mind-set.

Some economic analysts however discredit the claims of the government on petrol subsidy. Izielen Agbon (2011) analyzed thus:

> So let us conclude this basic economic exercise. If the true price of 38.2 per cent of our petrol supply from our local refinery is N34.36/liter and the remaining 61.8 per cent has a true price of N42.36 per liter, then the average true price is (0.382*34.36+0.618*42.36) or N39.30 per liter. The official price is N65 per liter and the true price of petrol in Nigeria is N39.30 per liter (even with our moribund refineries and imports). There is therefore no petrol subsidy. Rather, there is a high sales tax of 65 per cent at current prices of N65 per liter.

Even if various expert reports agree with the stand of Izielen, what cannot be denied is government's interference and regulation in petroleum which according to the anti-subsidy group created the opportunity for unending corruption and scams. However, in Libya, there is also government regulation in the petroleum sector but quite not corrupt like Nigeria's. While the need for petroleum subsidy remains, the Nigerian government needs to re-organize her bureaucracy and ensure effective reforms that will bring about healthy competition and encourage entrepreneurship in order to make the petroleum sector more viable.

It was also part of the sermon of the pro-subsidy removal lobby that petrol subsidy removal will check the corrupt practices of public

Chapter Four
Robert Madu & Shedrack Moguluwa, in
Jideofor Adibe (Ed.)
The Politics and Economics of Removing Subsidies on Petroleum Products in Nigeria
London & Abuja, Adonis & Abbey Publishers

93

officers and some 'money-bags' in the country. The authorities told the nation of a cabal that sits unperturbed on the profits of the fuel subsidy at the detriment of ordinary Nigerians. The question from the lips of the masses is whether the so called cabals are invincible and above the law?

There is indeed a cabal that is benefiting from the oil subsidy in which its unwholesome practices retard the development of the Nigerian downstream oil sub-sector. Illuminating on these corrupt practices during the House of Representatives' live debate on the oil subsidy removal on January 8, 2012, Hon. Godwin Ndudi Elumelu said that some of the oil importers faked import documents and claimed subsidy from government without importing any fuel (Obaka, 2012).

Considering the above submission in the face of the confession of the official witness of the Petroleum Product Pricing Regulatory Agency (PPPRA) in court in January 2013 that after

> Completing the documents checking cycle from the agencies and institutions – without verifying whether they are genuine or forged... no effort is made to authenticate the documents submitted throughout their document-checking procedures (Reginald-Stanley, 2013).

(Reginald-Stanley, 2013) suggests that the issue of corruption in the downstream sector of the Nigerian petroleum is an organized practice that benefits the top elites around the corridors of power.

Contrary to the argument that the bulk of the country's resources that should be channeled to infrastructural development and job-creation were swallowed up by the fuel subsidy funds, statistics have shown that corruption in public offices takes even more, leaving the country worse off for it. It is evident that the government has not been able to judiciously use the billions of dollars recovered from corrupt public officers in any infrastructural development.

The fact that the (PPPRA) can concur to an unverified transaction document and forward same to the Petroleum Support Fund for release of funds to the acclaimed marketers is enough fertile ground

Chapter Four | Robert Madu & Shedrack Moguluwa, in
Jideofor Adibe (Ed.)
The Politics and Economics of Removing Subsidies on Petroleum Products in Nigeria
London & Abuja, Adonis & Abbey Publishers

for the breeding of corrupt practices. This, from all indications, has nothing to do with the subsidy of petrol but a deliberate porous policy that empowers stealing of public funds in the country.

Periscoping corruption as the real bane of the Nigerian economy, one might begin to wonder what really is wrong with the establishment of refineries to reduce or totally end importation of petroleum products in Nigeria. The federal government of Nigeria had since 2002 been granting licenses for the establishment of private refineries to address the fuel importation problem in the country, but none of them had yet taken off. It could also be assumed to be the organized activities of the 'invincible' or 'untouchable' cabals since having a functional refinery in the country would not augur well with their 'business as usual' idiosyncrasy. An investigation by Daily Trust in fact revealed that "bureaucratic delays, lack of funding and technical incompetence have stalled the establishment of nine private refineries approved by the government" (Moyo, and Songwe, 2012).

Poverty

With the above scenario, the Nigerian Government decided for an immediate removal of petrol subsidy.

> The policy was brought by the government in the midst of insecurity, poverty, health challenges, economic instability and immeasurable challenges of nation building...Nigerians are suffering day and night; the United Nations Human Development Index puts Nigeria at 159 out of 177 countries, with 70.8 percent of the population living on less than $1 a day. Yet, the government was bent on making life more horrible. Nowhere is safe in Nigeria, schools are targets of bombers; houses are burnt down while places of worship has been turned into the house of death where the certainty that one will come out alive is not sure (Oluwasegun, 2013).

Infant mortality in Nigeria remains unacceptably high at 90.4 per 1,000 live births. In 2004, it was estimated that only 15 percent of the country's roads were paved, (Moyo, and Songwe, Op cit).

Chapter Four	Robert Madu & Shedrack Moguluwa, in
	Jideofor Adibe (Ed.)
	The Politics and Economics of Removing Subsidies on Petroleum Products in Nigeria
	London & Abuja, Adonis & Abbey Publishers

This is an opportunity for Nigerians to pause for a moment and reflect on what happened seven years ago. The subsidy that was removed on diesel and kerosene has not made any difference, the prices of these commodities still remain as high as ever and there is no indication they will drop. Yet, nothing came out of it, security, education, standard of living, nation building are all in a state of contemptible dismay and emotionally lacerating (Oluwasegun, Op cit).

Despite the cries and wailings of the Nigerian people, the government maintained that with the subsidy removal policy, about "N478bn is estimated to accrue to the Federal Government in 2012 from the policy" (Onuba, 2012). However, what the subsidy removal analysts forget to consider was that the weighty consequences of the policy on the already suffering Nigerians might leave them all dead before the supposed benefits of the subsidy removal policy will be felt.

Inconsistent Electric Power Supply

Nigeria has been burdened with epileptic power supply.   As Nwachukwu (2012)

> For a while now, Nigerians have been given all manner of excuses for the recent drop in electricity supply, despite assurances of improvement, but investigations have shown that it will take even a while longer before the country can experience some reprieve. This is because virtually all the generation plants are bugged down with one technical problem or the other. The reality is such that even if the Federal Government fulfils its pledge to revamp these faulty plants, Nigerians will still not have enough power to meet their industrial and domestic energy needs (Nwachukwu, 2012).

Vulnerable Nature of the Nigerian Economy

The nature of the Nigerian economy is such that the greater population of her work-force belongs to the low and middle income earners. Youth unemployment is still high and the nation is not

Chapter Four | Robert Madu & Shedrack Moguluwa, in
Jideofor Adibe (Ed.)
The Politics and Economics of Removing Subsidies on Petroleum Products in Nigeria
London & Abuja, Adonis & Abbey Publishers

witnessing any economic expansion to match the growing population. Also, with the depressed nature of the economy, many small and medium sized businesses find it difficult to access credit opportunity

The Nigerian Government seems to be so carried away and more interested in the supposed financial benefits accruable to it from subsidy removal that she ignored or rather played down on other grave consequences of her 'pre-mature' subsidy removal policy. It is obvious that when the effects of the unprepared removal of fuel subsidy descends in full force, the prices of commodities in the local industry will be hiked up, increasing the likelihood that imported products will have better footage in the competitive Nigerian market over the local industry.

This is really happening to an economy which also has an interest in the resuscitation of her local industries to compete with their foreign counterparts. In other words, the Nigerian government is indirectly setting a stumbling block for her internal policy for national economic growth and development. It therefore appears rather hypocritical encouraging the citizens to patronize locally made goods when the government's own action has set the pace for the triumph of the foreign industry above the local.

Prevailing Efforts of the Nigerian Government in the Face of these Challenges

The Nigerian Government in a bid to establish its ideals about the fuel subsidy removal policy established the Subsidy Reinvestment and Empowerment Programme (SURE-P) to oversee the modus operandi of the utilization of the money realized from the supposed fuel subsidy account.

> The programme was designed to act as receptacle for funds that accrued to the government from the partial removal of the subsidy on petroleum products. The funds are expected to be deployed to social service projects

Chapter Four | Robert Madu & Shedrack Moguluwa, in
Jideofor Adibe (Ed.)
The Politics and Economics of Removing Subsidies on Petroleum Products in Nigeria
London & Abuja, Adonis & Abbey Publishers

97

that will cushion the impacts of the increase in the pump price of petroleum products from N65 per litre to (N97) per litre, (This Day, 2013).

Ironically, within few months of the launch of this programme, the dust of corrupt practices around the subsidy fund started polluting the air in the nation. As Conscience Nigeria noted:

A Lagos High Court in Ikeja on Wednesday admitted as exhibit, a bundle of documents which an oil marketer, Rowaye Jubril, allegedly used to perpetrate a N963.7 million fuel subsidy fraud" (Conscience Nigeria, 2013).

Ameh (2002) also noted that the "House of Representatives' Adhoc Committee led by Hon. Farouk Lawal identified not just the problems of the industry but specifically, named individuals and companies who have diverted subsidies meant for the industry to private use".

It therefore came as a big shock to the Nigerian populace to learn that the highly respected legislator Hon. Farouk Lawal selected as Chairman of the Ad-hoc Committee investigating the fuel subsidy scam and his colleague Boniface Emenalo who served as the Secretary of the Committee got trapped in the same mire they came to clean up. As Nnochiri (2013 :) noted:

The Independent Corrupt Practices and other Related Offences Commission, ICPC, had in the charge, alleged that the embattled men (Farouk and Emenalo) demanded and collected bribe from the Chairman of Zenon Petroleum and Gas Ltd, Femi Otedola, as an inducement to remove the name of his company from the report of the House of Representatives Ad-hoc Committee on Monitoring of Fuel Subsidy Regime.

The Ribadu report provides another damning account of systemic sleaze, fraud and embezzlement in Nigeria's irredeemably corrupt oil sector. Covering 2002 to 2012, the report detailed how the country lost up to $29 billion in shady deals with oil majors; theft of an alarming 250,000 barrels of crude per day worth about $6.3 billion a year and how $183 million in signature bonuses from seven discretionary oil

Chapter Four | Robert Madu & Shedrack Moguluwa, in
Jideofor Adibe (Ed.)
The Politics and Economics of Removing Subsidies on Petroleum Products in Nigeria
London & Abuja, Adonis & Abbey Publishers

98

prospecting licences issued between 2008 and 2011 were not accounted for. Three of the licences were said to have been issued under current Petroleum Minister, Diezani Alison-Madueke, (Onuba, 2012).

Today, the problem with the petroleum industry is largely lack of decency and political will on the part of the government to deal with those who have already been identified as having corruptly enriched themselves with funds meant for the industry. Majority of these people are suspected to be political associates of those in power. President Jonathan has in fact openly been hobnobbing with some of the key persons indicted by the reports, (Ameh, Op cit).

The Imperativeness of Subsidy to the Sustainability of the Nigerian Economy

In a global sense and comparative terms, countries and regions have applied subsidy to their members and citizens over the years. For example, the financial bail-out of the Republic of Ireland by Sweden; that of Greece undertaken by Germany and France and that of Cyprus and Spain carried out by the European Commission. All these are kinds of subsidy but culturally were applied in different circumstances to meet the needs of the citizens.

It is of a great necessity in this discourse to adjudge that the subsidy removal will leave the Nigerian economy in a more depressed state than it met it considering the complexities and peculiarities surrounding the Nigerian polity. As Balouga (2012) noted:

> After many years of control and uncertainty surrounding the sale and purchase of petroleum products in Nigeria, the government is now deciding to emulate other developing and developed nations to fully privatize and liberalize the country's downstream sector which is managed by the National Petroleum Corporation (NNPC) on behalf of the government.

Chapter Four | Robert Madu & Shedrack Moguluwa, in
Jideofor Adibe (Ed.)
The Politics and Economics of Removing Subsidies on Petroleum Products in Nigeria
London & Abuja, Adonis & Abbey Publishers

99

It is not just wise enough to embrace total capitalism of deregulating the petroleum down-stream sector of Nigeria in emulation of the developed western countries whose economies are livelier than what is obtainable in Nigeria. Citing Dansie, et. al (2010), Abang, et. al (2012) wrote:

> Subsidy removal may however sound nice, polite and beneficial, it is essential that there be a critical evaluation of the proposed mechanism for petroleum scarcity alleviation. Lessons from developed countries where this commodity has become readily available at all time reveals that subsidy removal is not a significant measure responsible for such achievement. China for instance still gives subsidy to her citizen on petroleum product and the fact available today showed that it is still among the first five economy of the world that could boast of adequate petroleum product and economic stability.

If anything, the Nigerian government should have borrowed a leaf from the Chinese government whose population is far more than Nigeria's. A government is established for the benefits and well-being of the people and not the people for the government. On this notion therefore, every policy of the government should be evaluated on the basis of its proximity to the interest of the people. As Oluwamayowa, (2011) puts it:

> Briefly, I would like to tell a non-fictional story of countries enjoying fuel subsidy. The first country being Libya, In Libya, the pump price of petrol is about N16 per liter, with the global price being N156 per liter. This is an amazing N140 subsidy if the fuel was imported, yet Libya, under a Dictator was the country with the highest human development index in Africa, still Libya had no external debt. The second country is Egypt, a non-oil producing state. Pump price in Egypt is about N47 per liter, with great social economic packages.

Chapter Four | Robert Madu & Shedrack Moguluwa, in
Jideofor Adibe (Ed.)
The Politics and Economics of Removing Subsidies on Petroleum Products in Nigeria
London & Abuja, Adonis & Abbey Publishers

100

Consequences of Subsidy Removal in a Depressed Nigerian Economy

The removal of fuel subsidy will only amount to sacrificing the same people that are supposed to be saved by the government. It is worth mentioning that the present situation of the Nigerian citizenry should have been considered first before bringing the future proceeds of the deregulation gains into the picture of the premature policy on them. Indeed there are other nations (even in Africa) like Ghana that have removed fuel subsidies but never was it done anywhere else that the subsidy was removed unprepared and in a crash-like and haphazard manner like Nigeria did on January 1 2012.

Efforts of the government should have been geared towards making preparations that would cushion the short term effects of the removal of the petroleum subsidy before its actual removal. This is because the de-subsidization will affect all facets of the peoples' lives and the nation will live with such consequences on daily basis. As Abang, et. al, (Op cit) noted:

> Previous records have shown that any increase in the price of petroleum product automatically lead to an increase in the cost of transportation of both raw and finished product which in effect will lead to an increase in the cost of production for end user.

The effects occur like chain-reaction and have daily implications for many of the people within the bottom of the economy. Also, a related argument is the structure of Nigerian economy, where energy (electricity) solely built around petrol (fuel), has made the product the prime determinant of the prices of other goods and services in the country. This means therefore that a hike in the price of fuel will likely result in the further devaluation of the Naira (Nigerian currency). There is a high probability that this will come since the rise in the pump price of petrol will result in depreciating the purchasing power of the Naira. When this comes to be, the end consequence will

Chapter Four   Robert Madu & Shedrack Moguluwa, in
Jideofor Adibe (Ed.)
The Politics and Economics of Removing Subsidies on Petroleum Products in Nigeria
London & Abuja, Adonis & Abbey Publishers

101

be an increase in the poverty level of the citizenry, which still stalls or completely floors the government's policy on poverty alleviation.

No matter how juicy the promises and baits of the subsidy removal policy of the Nigerian Government appear, the truth remains that there are a lot of other measures the government would have adopted even without the instant removal of fuel subsidy that would have bettered the depressed Nigerian economy significantly. Needless to say that the state of insecurity, criminality and lawlessness in the country has a direct bearing on the inherent high cost of living in the face of a mass unemployment situation despite all the noise about poverty alleviation and job-creation by the Nigerian Government.

Petroleum Subsidy SWOT Analysis

	S (Strengths)	W (Weaknesses)	O (Opportunities)	T (Threats)
1	Cheap and affordable price of petroleum products	Government incurs the huge debt of subsidizing petroleum products	Poverty reduction as a result of affordable cost of energy comparatively to the cost of other commodities	Corrupt public officers hide under the cloak of subsidy funds for shoddy deals
2	Affordable price of other goods and services in the country	Encourages petrochemical dependent economy in Nigeria	Higher buying power of the Naira brings increase in demand hence will contribute to economic growth	Non-competitiveness of the petroleum down-stream sector hampers job creation from the sector
3	Boost to the local industry to stand strong in competition with international contemporaries	Discourages private investors (capitalists) in the down-stream sector of the Nigerian petroleum industry	Encourages the growth of Small Scale Enterprises (SSE) which the middle and low income earners dominate	Incentives to agriculture and other economic sectors are threatened by the subsidy fund
4	Sustainability and increase in the value of the Naira (Nigerian currency)	Subsidy swallows the fund that should have been channelled into other developmental	Ensures popularity of the government on the side of the citizenry as that has been proven to be the	Scarcity of petroleum product reduces GDP and

Chapter Four | Robert Madu & Shedrack Moguluwa, in
Jideofor Adibe (Ed.)
The Politics and Economics of Removing Subsidies on Petroleum Products in Nigeria
London & Abuja, Adonis & Abbey Publishers

		projects in the long run	yearning of the ordinary Nigerian	threatens economic growth and sustainability
5	Increase in the Gross Domestic Product (GDP) of Nigeria	Incessant scarcity of petroleum products resulting to stagnation of economic activities	Will cushion the effects of unemployment and poverty on the citizens of Nigeria	Leaves the Nigerian economy with huge external debt
6	Higher and more improved standard of living in Nigeria	Disproportionally benefits the rich who consumes much of the subsidized products at the detriment of the poor	Will to a very large extent check inflation in the economy of Nigeria	Makes the country a puppet in the hands of her donor nations for aids and grants

Conclusions

While the argument for subsidy removal could be appreciated, it appears elitist in nature, lopsided and requires more time for the Nigerian society to gain more stability and sustainability. For example, attainment of middle class economy, reduction of infant mortality rate, increase in the life span of the average Nigerian, high reduction in the level of unemployment, infrastructural development, especially in the Nigerian rural areas, introduction of welfare packages to alleviate the impact of high levels of poverty and achieving sustainable rebranding of the image of the Nigerian Government as a trustworthy entity before the citizenry and international community should have come before the attempt to fully remove the subsidy on fuel.

The recent report of the National Bureau of statistics (NBS) as reported by Sub air (2012) indicated that the;

> ...percentage of Nigerians living in absolute poverty – those who can afford only the bare essentials of food, shelter and clothing – rose to 60.9 per cent in 2010, compared with 54.7 per cent in 2004...Although Nigeria's economy is projected to continue growing, poverty is likely to get worse as the gap between the rich and the poor continues to widen.

Chapter Four | Robert Madu & Shedrack Moguluwa, in
Jideofor Adibe (Ed.)
The Politics and Economics of Removing Subsidies on Petroleum Products in Nigeria
London & Abuja, Adonis & Abbey Publishers

Recommendations

From the foregoing, we will make the following recommendations:

1. The Nigerian government must encourage investment in alternative energy sources.
2. Serious reforms are needed in the Nigerian petroleum sector to encourage healthy competition, entrepreneurial ingenuity and innovations in the sector.
3. The Nigerian government must encourage anti-corruption campaigns, especially in the petroleum sector.
4. The government should establish social welfare packages that will aid in the alleviation of poverty among the masses.

References

Abang, I.S., Elufisan, T.O., and Okwubunne, A.C. (2012), 'Linear Function Application: Enlightenment to the Impact of Fuel Subsidy Removal in Nigeria', American Journal of Economics, Scientific and Academic Publishing Co. Available at www.article.sapub.org/10 (Accessed: 21 March 2013).

Aliede, E.J. (2010), 'Corporate social Responsibility (CSR) as a Mitigative Strategy for the Sustainable Resolution of the Protracted Crisis in the Niger Delta' in B. Madu (ed.), *Public Relations and Media Communications: African Perspective*, Enugu: Immaculate Publications Limited.

Ameh, C.G. (2012), 'NLC Warns Jonathan: Total Removal of Fuel Subsidy will Collapse your Government', *Daily Post*, November 17, Available at www.dailypost.com.ng (Accessed: 21 March 2013).

BBC News Africa (2012) 'Nigerians Living in Poverty Rise to nearly 61%', Available at http://www.m.bbc.co.uk/news/world-africa (Accessed: 21 March 2013).

Chapter Four | Robert Madu & Shedrack Moguluwa, in
Jideofor Adibe (Ed.)
The Politics and Economics of Removing Subsidies on Petroleum Products in Nigeria
London & Abuja, Adonis & Abbey Publishers

104

Balouga, J. (2012), 'The Political Economy of Oil subsidy in Nigeria', *International Association for Energy Economics*, Available at www.iaee.org/en/publications/news (Accessed: 21 March 2013).

Conscience Nigeria, (2013), 'Group Urges Nigeria to Remove Fuel subsidy, Build Coastal Refineries', *Premium Times*, January 13, Available at (www.premiumtimesng.com/news/114970) (Accessed: 21 March 2013).

Doreo Partners (2013) 'Alarming Food Importation Rate in Nigeria', Available at (www.doreopartners.com) (Accessed: 21 March 2013).

Gbola Subair, (2012), 'Nigeria's Poverty Level Rises, Hits 71.5%, Sokoto, Niger, top List of Poorest States' *Nigerian Tribune*, Thursday, March 21, 2013, Available at www.nigeriantribune.com .ng/index.php/fro (Accessed: 21 March 2013).

Izielen, A. (2011), 'The Cost of One Litre of Petrol', *The Guardian*, Nigeria, Available at www.ngrguardiannews.com/index.php?opti on (Accessed: 25 March 2013).

Mohammed Hamisu (2013), 'Nigeria: Oil Refinery Projects Stall', Daily Trust, 14 January, Available at www.m.allafrica.com/story (Accessed: 21 March 2013.

Moyo, N., and Songwe, V. (2012), 'Removal of Fuel subsidies in Nigeria: An Economic Necessity and Political Dilemma', Global Economy and Development, *Africa Growth Initiative*, Available at www.brookings.edu/research/opinion (Accessed: 21 March 2013)

Nnochiri, I. (2013), 'Subsidy Scam: The Futile Bid to Save Farouk and Emenalo from Detention, Available at www.vanguardngr.com (Accessed: 21 March 2013).

Nwachukwu, C. (2012), 'Steady Power supply: Nigerians to Wait Longer', *Vanguard* December 24, Available at www.vanguardngr.c om/ 2012/12/steadypowersupply (Accessed: 21 March 2013).

Obaka, A.I. (2012), 'Issues in Oil Subsidy Removal', *The Punch*, January, 23, Available at www.punchng.com (Accessed: 21 March 2013).

Chapter Four	Robert Madu & Shedrack Moguluwa, in
	Jideofor Adibe (Ed.)
	The Politics and Economics of Removing Subsidies on Petroleum Products in Nigeria
	London & Abuja, Adonis & Abbey Publishers

Oluwamayowa, T. (2011), 'The Positives of Oil subsidy Removal in Nigeria', Available at www.saharareporters.com (Accessed: 21 March 2013).

Oluwasegun, O.D. (2013), 'Prospective View of Fuel Subsidy Removal', *Information Nigeria*, January 8, Available at www.informationng.com (Accessed: 21 March 2013).

Onuba, I. (2012), 'Subsidy Removal: Nigerians won't Suffer in Vain – Okonjo-Iweala', *The Punch*, January 8, Available at www.punchng.com (Accessed: 21 March 2013).

Okigbo, Pius (2013), 'How IBB Wasted $12bn Oil Windfall', Available at www.nairaland.com (Accessed: 26 March 2013).

Reginald-Stanely (2013), 'PPPRA does not Authenticate Documents of Fuel Importers before Processing their Subsidy Claims: Its Officials Tell Lagos Court', Available at www.saharareporters.co m (Accessed: 21 March 2013).

The Nation Archive, (2007), 'Nigeria: Wretched Excess', Available at www.thenationonline.ng (Accessed: 26 March 2013).

Thisday, (2013), 'Probing Subsidy Re-investment Programme', Thisday, Friday 22, March, Available at www.thisdaylive.com/arti cles/probing (Accessed: 21 March 2013).

Chapter Four | Robert Madu & Shedrack Moguluwa, in
Jideofor Adibe (Ed.)
The Politics and Economics of Removing Subsidies on Petroleum Products in Nigeria
London & Abuja, Adonis & Abbey Publishers

CHAPTER FIVE

Petroleum Products subsidies in Nigeria: Economic and Global Perspectives

Emmanuel Ojameruaye

This chapter examines the economic and global perspectives of the petroleum products subsidy debate. We begin by defining the concept of subsidy and measurement of petroleum products subsidy. Next we examine the impact of subsidy on the petroleum products market and the socio-economic arguments for and against petroleum products subsidy in Nigeria. Thereafter we examine the issue of petroleum products subsidies from a global perspective. We conclude the chapter with recommendations on how to resolve the petroleum products subsidy conundrum in Nigeria.

The Definition and Measurement of Petroleum Products Subsidy in Nigeria

A subsidy (or subvention) is an amount of money paid by government to suppliers of a product or service to enable them to sell it to final consumers at a government-determined price that is less than the actual supply cost. For instance, the federal government (FG) provides subvention (subsidy) to federal universities so that they can charge tuition (demand price of university education paid by students) that is less than the actual cost of providing university education (i.e., supply cost). Thus, if a federal university has 10,000 registered students and if its total budget is N5 billion, its cost of providing education (supply cost) will be N500, 000 per student a year (N5billion/10,000). However, if the FG decides to peg tuition at N200, 000 per student, then it must provide a subsidy (subvention) of N300, 000 per student (i.e. N3 billion for all 10,000 students) to enable

Chapter Five | Emmanuel Ojameruaye, in
Jideofor Adibe (Ed.)
The Politics and Economics of Removing Subsidies on Petroleum Products in Nigeria
London & Abuja, Adonis & Abbey Publishers

107

the university provide its services (education) to the students at N200, 000 tuition.

The government of Nigeria subsidizes imported petroleum products[13] because it has decided to fix the prices of these products at levels lower than the cost of importation (and production) and delivery of the products to final consumers at the filling stations. For instance, if the cost of importation (landing cost) of petrol is N132 per liter and distribution/marketing cost is N15 per liter, then the supply cost will be N147 per liter. If the government decides to fix the pump price of petrol at N97 per liter, then it must pay a subsidy of N50 per liter (i.e. N147 – N97) to the suppliers of imported petrol (i.e. importers) so that they can deliver imported petrol to consumers at N97 per liter instead of the actual supply cost of N147 per liter. If the suppliers import and deliver 16 million liters of petrol a day to consumers (i.e. approximately 100,000 barrels a day), then the government will have to pay 16 million x N50 = N800 million a day or N212 billion a year to importers/suppliers of petrol as price subsidy in order to keep the pump price at N97 per liter.

In addition to subsidizing imported petroleum products, the government also subsidizes petroleum products produced by NNPC-owned local refineries because the supply cost (production + distribution cost) of locally produced products is also more than government-fixed pump price. For instance, if the cost of producing petrol by the local refineries is N92 per liter and the distribution/marketing cost is also N15 per liter, then the supply cost will be N107 per liter, and the subsidy will be N10 per liter (N107 – N97). Thus, if the refineries are delivering an average of 24 million liters of petrol (about 150,000 barrels) a day to consumers at N97 per liter, then the government must provide 24 m x N10 = N240 million a day or N63.6 billion a year to the refineries as subsidy or subvention.

---

[13] The products include petrol (otherwise known as gasoline or motor premium spirit or PMS), household kerosene (HHK), liquefied petroleum gas (LPG) and diesel or automotive gas oil (AGO).

Chapter Five	Emmanuel Ojameruaye, in
	Jideofor Adibe (Ed.)
	The Politics and Economics of Removing Subsidies on Petroleum Products in Nigeria
	London & Abuja, Adonis & Abbey Publishers

However, if the actual production cost of petrol by the local refineries is N82 or less and the distribution cost is N15 per liter, then the supply cost will be N97 or less, and the government will not have to pay subsidy or subvention to the refineries (NNPC).

Data on the actual production cost of petroleum products by the local refineries are not readily available to enable us to determine the amount of subsidy government is (or should be) providing to the local refineries. However, the unit production cost depends largely on the price at which crude oil is delivered to the local refineries, i.e. the price local refineries pay for the crude oil they use. It can be argued that NNPC can afford to deliver crude oil to the refineries at less than N82 per liter given the low cost of producing crude oil in Nigeria vis-à-vis the export (spot market) price of Nigerian crude oil. Assuming the unit production cost of crude oil is $10 per barrel (compared to spot market price of about $100 per barrel) which translates to about 6.2 cents or about N10 per liter, one can assume that the true cost of producing petrol after allowing for refining cost cannot be more than N20 per liter.

If crude oil is delivered to the refineries at less than the international or export (spot market) price of crude oil (say $100 per barrel = 62 cents or N100 per liter), then there is an implicit subsidy in terms of income forgone. If the local refineries are required to pay the spot market price (say N100 per liter) for the crude oil they receive from NNPC, then their unit production cost of petrol will be anywhere from about N105 to N120 per liter depending on their refining cost, and this amount will be close to but less than the unit cost of imported petrol. In this case, the FG will pay the difference between NNPC's unit cost and the fixed pump price as subsidy to the refineries (NNPC). Therefore, the existence and level of subsidy on locally produced petroleum products depends on the price at which crude oil is delivered to the local refineries. Unfortunately we have no information on the price at which crude oil is currently delivered to the local refineries. What we know is that in 2002, the price at which crude oil was delivered to the local refineries was increased from

Chapter Five | Emmanuel Ojameruaye, in
Jideofor Adibe (Ed.)
The Politics and Economics of Removing Subsidies on Petroleum Products in Nigeria
London & Abuja, Adonis & Abbey Publishers

$9.50 per barrel to $18.00 per barrel. If the price has not changed since then, then we can say that crude oil is still being delivered to local refineries at $18 per barrel which is almost N18 per liter ($18 xN160/159 liter). The Group Managing Director (GMD) of NNPC told the Senate Committee investigating the fuel subsidy issue in 2011 that it cost between $4 and $5 to refine a barrel of crude oil depending on the exchange rate of the naira (*Vanguard*, December 4, 2011). This translates to about N5 to refine a liter of crude oil into petrol and other products. The GMD further stated that NNPC

> ...collects subsidy on locally refined products but it is less than what we collect on imported product. But since we do not include trade marking and other charges on imported products, it costs N11.87 less on locally refined products and N11.87 more on imported ones (*Vanguard*, December 4, 2011).

The government's claim that it is subsidizing imported fuel cannot be disputed given the fact that the spot market price of petroleum products is far higher than the pump price at Nigeria's filling stations. For instance, at the height of the fuel subsidy debate in 2011 the spot market price (North European/Rotterdam, f.o.b.) for regular petrol unleaded was $92.7 per barrel or about N93 per liter in October 2011, which was much higher than the pump price of N65 per liter in Nigeria. However, according to the PPPRA's pricing template in October 2011, the import price (c.i.f.) for petrol was N117.78 per liter while to the supply cost was N142.13 per liter, thus the subsidy rate was N77.13 per liter (N142.13 – N65). Estimating the actual amount paid out as subsidy in Nigeria is complicated because the different agencies of government sometimes provide different and conflicting figures. For instance, the PPPRA claimed that the federal government spent N3.655 trillion on subsidy between 2006 and October 2011 including N1.54 trillion during the 10 months of 2011.On the other hand, the budget for subsidy for 2011 was N245 billion while the Federal Ministry of Finance (FMF) stated that the government paid N1.54trillion (which is about N1.3 trillion in excess of the budgeted amount) as subsidy to importers of fuel between

Chapter Five | Emmanuel Ojameruaye, in
Jideofor Adibe (Ed.)
The Politics and Economics of Removing Subsidies on Petroleum Products in Nigeria
London & Abuja, Adonis & Abbey Publishers

January and October 2011. NNPC claimed that the excess amount paid as subsidy was N192.5 billion which was a far cry from the N1.3 claimed by the FMF. In fact, the Executive Director of PPPRA, Reginald Stanley, told the Senate Committee that the gross amount spent on fuel subsidy from 2006 to September 2001 year was N3.655trillion which contradicted the N1.426 trillion submitted by the NNPC as subsidy as of August 2011. He also refuted the N450 billion kerosene subsidies owed the NNPC by the FG (*Vanguard*, December 4, 2011).

The excess amount paid as subsidy cannot be justified on the basis of the quantity of petroleum products imported, the import price of the imported products and prevailing exchange rate. These were strong indications that grand corruption accounted for the astronomical increase in subsidies paid out in 2011 (and later in 2012). In a fact sheet presented at a meeting between President Jonathan and leaders of some political parties, the Minister of Finance stated that:

> The large increase observed in 2011 is as a result of (i) increased crude oil price from US $81.25 per barrel(pb) to $US 113.98pb; (b) exchange rate movements; (c) larger volumes consume(about 35m liters per day); and (iv) N150billion of kerosene carried over from 2009 and 2010 (*The Nation*, December 5, 2011).

The dubious list of importers who received the subsidies in 2011 is an indication of the fact that corruption accounted for a substantial part of the increase in the amount of fuel subsidy. The House of Representative Ad Hoc Committee on the Fuel Subsidy Report revealed that the astronomical increase is subsidy payment was due to corrupt practices.

Another problem with estimating the actual level of fuel subsidy in Nigeria is the fact that the pricing template which PPPRA uses in computing the amount of subsidy appears biased in favor of importers. Makwe (2006) has observed that the PPPRA pricing template overstates the level of subsidy because: a) the import prices used in the template are generally higher than the international (spot)

Chapter Five | Emmanuel Ojameruaye, in
Jideofor Adibe (Ed.)
The Politics and Economics of Removing Subsidies on Petroleum Products in Nigeria
London & Abuja, Adonis & Abbey Publishers

market prices (after adjusting for the difference between "fob" and "cif"); b) the template provides for excessive financing charges, port and storage charges and margins for importers, transporters, dealers, distributors and retailers; and c) the pricing is not competitive.. Thus, if the subsidy is removed and the current template is used to determine pump price of petroleum products, consumers will be unduly penalized and the prices of petroleum products will most likely be higher than "free market" prices.

The Impact of Subsidy on the Petroleum Products Market

Subsidy affects the demand for and supply of petroleum products in Nigeria in a significant manner. To illustrate this, we will use the demand and supply analysis in economics. In figure 1 below, let the line Do represent the demand curve while the line So represents the supply curve for petrol. Do is negatively sloped meaning that the higher the price of petrol, the more the quantity of petrol that will be demanded or consumed. On the other hand, the supply curve is positively sloped which means that the higher the price of petrol, the more quantity will be supplied, either by the local refineries or importers. Assuming there is no subsidy paid by government to suppliers, the market will be cleared (i.e. be at equilibrium) at point Eo where Po = N100 per liter, and Qo = 300k barrels a day. If government decides to fix the price of petrol at N65 per liter (P1), the quantity demanded will increase to Q1 = 400k barrels a day. To supply this quantity of petrol, suppliers will require a unit price of P1 = N140 per liter. Therefore the government must pay the difference between the fixed price (N65) and the price suppliers want (N140). This difference N140 – N65 = N75 per liter is the subsidy on each liter of petrol supplied to the market and purchased/consumed by motorists and other users of petrol. Thus, the effect of subsidy is to make the selling price of petrol (and any product that is subsidized) to consumers lower than the true supply cost (market equilibrium price). This means a shift of the supply curve (So) downward by the

Chapter Five | Emmanuel Ojameruaye, in
Jideofor Adibe (Ed.)
The Politics and Economics of Removing Subsidies on Petroleum Products in Nigeria
London & Abuja, Adonis & Abbey Publishers

amount of the subsidy to S1 which means an effective increase in supply. A subsidy is therefore the equivalent of a negative tax, since consumption tax has the opposite effect.

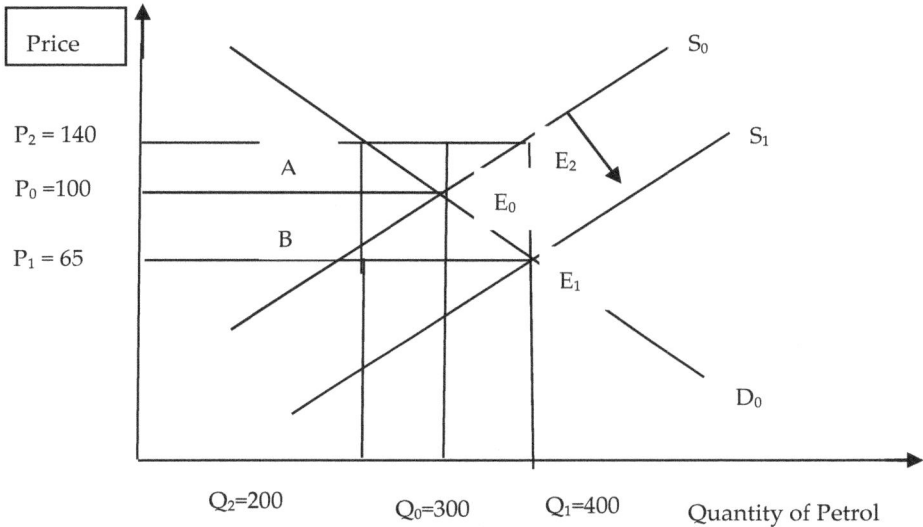

Fig. 1: Chart showing the effect of subsidy on petrol

In the above chart, P1 = 65 is the price paid by consumers after the subsidy is created. P2 = 140 is the price received by the suppliers which is the price paid by consumers (65) plus the subsidy (75). Note that before the introduction of subsidy, the equilibrium price was Po = 100. Thus, the effects of the subsidy are to lower the price consumers pay and increase the price received by suppliers. The benefit of the subsidy is shared by the consumers and suppliers in a proportion that depends upon the relative slopes (elasticity) of the demand and supply functions. The chart shows that both consumers and suppliers benefit as a result of the subsidy, but the government incurs the cost. How does the benefit received by consumers and suppliers compare with the cost incurred by government? In the chart, the cost of the subsidy to the government is represented by area of the rectangle P2E2E1P1 (= 75 x 400 = 30,000 units); the benefit

Chapter Five | Emmanuel Ojameruaye, in
Jideofor Adibe (Ed.)
The Politics and Economics of Removing Subsidies on Petroleum Products in Nigeria
London & Abuja, Adonis & Abbey Publishers

received by consumers is represented by the area of the trapezoid PoEoE1P1 (=35 x300 + 0.5x35x100 = 12,250 units); and the benefits received by suppliers is represented by the area of the trapezoid P2E2EoPo (= 40x300 + 0.5x40x100 = 14,000 units). The chart shows that the area of the rectangle P2E2E1P1 (30,000 units) representing the cost of subsidy is greater than the sum of the areas of the two trapezoid (12,250 + 14,000 = 26,250 units) meaning that the net cost is positive which is the same as net benefit being negative or the benefit/cost ratio being less than one. Thus, theoretically, economists say that subsidies are Pareto inefficient because they cost more than they deliver in benefits.[14]

In the above illustration, the suppliers benefit more than the consumers but this need not be the case always. What is clear is that the sum of the benefits is always lower than the cost incurred by government. The difference between the cost and benefits is called deadweight loss. In this example it is 30,000 – 26,250 = 3,750 units. We have also lumped all consumers (industrial, rich people, middle income people and poor people) together. However, for practical and policy purposes, it is necessary to disaggregate the benefits enjoyed by each group of consumers in order to study the effect of removal of subsidy on each group. For instance, the Minister of Finance has argued that the current fuel subsidy benefit rich people who own many cars more than the poor. However, while this may be true in absolute terms, it may not be true in relative terms if the share of expenses on petroleum products in the budget of poor families is higher than those of the rich or if the rich can easily compensate for the loss of income due to inflation arising from the removal of subsidy while the poor can hardly do so.

---

[14] A situation or an allocation is said to be Pareto inefficient if at least one person can be made better off without making anyone else worse off or if someone can be made better off even after compensating those made worse off. In the case of subsidies, it means that eliminating subsidies can make government better off after compensating for the losses of consumers and suppliers.

Chapter Five | Emmanuel Ojameruaye, in
Jideofor Adibe (Ed.)
The Politics and Economics of Removing Subsidies on Petroleum Products in Nigeria
London & Abuja, Adonis & Abbey Publishers

The fact that subsidies are Pareto inefficient does not necessarily justify their removal or reduction. As long as a government imposes consumption tax (such as value added tax or sales tax on some commodities), it can be argued that the government should also subsidize certain commodities, especially those that are "critical" to the economy and those that benefit the poor more than the rich. Thus, while many economists will argue that the elimination of fuel subsidy is good for the economy and is a "win-win" or "no lose" situation at least from a theoretical point of view, others have embraced the political and social realities that either justify fuel subsidies or make their elimination impracticable, at least in the short to medium terms (Victor, 2009). It is therefore important to examine the wider economic, political and social case for and against fuel subsidies.

The Case for Petroleum Products Subsidy in Nigeria

Some of the reasons that have been adduced for fuel subsidies in Nigeria and some other countries include the following:

a) Fuel subsides are necessary to control price inflation and prevent a decline in the real income and living standards of consumers, especially lower income households. Removal or reduction of subsidy will not only increase the prices of petroleum products but also the prices of most other products and services in the economy. This in turn may lead to galloping inflation and further pauperization of the poor in a country where over 60% of the people are already living below the poverty line. In fact, a look at the components Nigeria's consumer price index easily shows that the price of energy is one of the key drivers of inflation in the country because the "accommodation, fuel and light" price index has generally been the higher than the composite price index and those of other components except "household goods and other purchases". However, evidence shows that the inflationary

Chapter Five | Emmanuel Ojameruaye, in
Jideofor Adibe (Ed.)
The Politics and Economics of Removing Subsidies on Petroleum Products in Nigeria
London & Abuja, Adonis & Abbey Publishers

115

impact of a moderate in the price of petroleum products is not high as feared. For instance, the CBN Half Year Report for Jan-June 2012 showed that

> The core inflation, i.e. all items less farm produce, on a year-on-year basis stood at 15.2% by the end of June 2012 compared to 11.5% at the end of June 2011. On a 12-montl moving average basis, core inflation was 12.7% compared to 12.1%. The rise in core inflation was accounted for largely by the effect of the partial removal of subsidy in the pump price of PMS (CBN, 2012)

b) To reduce the cost of production and help to stimulate economic growth by increasing long-run aggregate supply. Since energy cost accounts for a significant part of the cost of production, especially in the industrial sector, a significant increase in the price of petroleum products arising from the removal or reduction in fuel subsidy, will lead to an escalation of the prices of commodities in the economy, which will lead to a reduction in local demand and make exports less competitive. This could lead to a reduction in production, reduction in the demand for labor, increase in unemployment, and, consequently, economic recession. Thus, in order to stimulate economic growth, governments are very cautious in eliminating or reducing existing subsidies. In fact, subsidies are sometimes retained because of the need to prevent massive lay-offs or to boost employment, especially during times of high unemployment.

c) To smoothen the process of long term structural change or transformation in certain industries and prevent a decline in the production of some agricultural crops such as cotton. Elimination or significant reduction of subsidies can led to a rapid decline or collapse of some ailing industries or those going through structural changes and adaptation. Thus, in order to prevent such as rapid decline, subsidies may be retained or reduced gradually.

Chapter Five | Emmanuel Ojameruaye, in
Jideofor Adibe (Ed.)
The Politics and Economics of Removing Subsidies on Petroleum Products in Nigeria
London & Abuja, Adonis & Abbey Publishers

d) To encourage the provision and consumption of "merit" goods and services which generate positive externalities (increased social benefits). Merit goods and services are those that are provided free or at a subsidized rate by government for the benefit of the entire society because they would be under-provided if left to the market forces or private enterprise. Examples of merit goods are healthcare, education, security and museum. On the other hand, demerit goods and services are those that harm the consumer and have negative externalities. Examples include cigarettes, alcohol, drugs and prostitution. Under-consumption of merit goods can lead to market failure which can lead to a reduction in social welfare. Thus, it is necessary to subsidize some merit goods or services to prevent under-consumption or weak demand. Although petroleum products are not merit goods, they do have some of the qualities of merit goods. For instance, the removal of subsidy on kerosene may force the poor to resort to greater use of firewood for cooking which will in turn increase deforestation and greenhouse gases. Therefore, the elimination or reduction of fuel subsidy must be approached with caution to avoid negative externalities.

e) To prevent industrial action, protests and riots that can lead to political instability. Labor unions, non-governmental organizations and the poor generally try to resist any attempt to reduce or eliminate subsidies. In some cases, it can lead to uncontrolled strikes, protests and unrests that can lead to political instability and changes in government. In some less developed countries, there have been several cases of "food riots" due to the removal of subsidies on some food items, especially imported food items, which has resulted in high prices of such food items. Thus, the government must weigh the risks of political instability that could arise from the removal or reduction of fuel subsidy.

Chapter Five | Emmanuel Ojameruaye, in
Jideofor Adibe (Ed.)
The Politics and Economics of Removing Subsidies on Petroleum Products in Nigeria
London & Abuja, Adonis & Abbey Publishers

The Case against Petroleum Products Subsidy in Nigeria

There are also several counter arguments against subsidies, especially on the grounds of economic efficiency and equity. The fact that subsidies are Pareto inefficient is the theoretical case (necessary condition) against subsidies, but it does not mean that all subsidies must be eliminated. Other arguments against petroleum products subsidy in Nigeria include the following:

a) Market distortions: Free market economists argue that subsidies distort the free market mechanism and can worsen the distribution or allocation of resources. For instance, subsidy on some agricultural commodities like cotton in industrialized countries distort free trade in those commodities and severely curtail exports from less developed countries (LDC). Similarly, import subsidy on petrol by in a LDC may discourage domestic production of petrol and lead to misallocation of an increasing amount of scarce foreign exchange for importation of petrol.

b) Subjective decision: The decision to subsidize a commodity and which groups of suppliers to receive a subsidy can be arbitrary and subjective. It can also be prone to corruption. More often than not, there are no objective standards to determine which commodities should be subsidized, the level of subsidy and which suppliers should receive the subsidy. Why subsidize petrol instead of electricity, or fertilizers instead of rice, or education instead of health, etc? Why should subsidy be paid to suppliers of imported petrol and not to local refineries (local producers)? Are the suppliers selected through a competitive process or are they "friends" and "supporters" of the government and decision-makers? Is the amount of subsidy paid determined through a competitive process that ensures value-for-money, or do the suppliers influence the amount of subsidy they receive and pay "kick-

Chapter Five | Emmanuel Ojameruaye, in
Jideofor Adibe (Ed.)
The Politics and Economics of Removing Subsidies on Petroleum Products in Nigeria
London & Abuja, Adonis & Abbey Publishers

backs" to decision makers? For instance, it has been alleged that that petroleum pricing template which the PPPRA uses in determining the subsidy on imported fuel is biased in favor of the suppliers because it allows for very "generous" charges and margins and that some of the cost elements can be reduced through competition. For instance, the pricing template for petrol (PMS) for November 2011 includes a financing charge of N2.49 per liter and storage cost of N3 per liter in the landing cost, as well as a bridging fund charge N5.85 per liter, retailers margin of N4.60 per liter, transporters margin of N2.99 per liter, dealers margin N1.75 per liter, Marine Transport Average (MTA) cost of N0.15 per liter) and Administrative charge N0.15 per liter, all of which appear too generous.

c) Risk of fraud and corruption: Subsidies are susceptible to corruption and the ever-present risk of fraud, especially when allocating subsidy payments. For instance, the delay in the reimbursement of subsidies to importers of fuel has created incentives for the importers to induce payment (U4 Anti-Corruption Resource Centre, 2009). There are also several reports of high-profit rackets and "round-tripping" of imported fuel and fuel produced by local refineries (Nuhu-Koko, 2008).

d) Cost of subsidies: The cost of subsidy can grow rapidly and become unsustainable in the face of increasing demand for the subsidized product. If not brought under control, subsidies on a few products can crowd out expenditures or investments in other vital sectors such as roads, electricity, public water supply, electricity, education and health. The federal government is arguing that Nigeria is close to such a point in the sense that it paid a whopping sum of N1.3 trillion between January and October 2011 to suppliers of petroleum products. This is the equivalent of giving every Nigerian about N8,000 a year or N40,000 to a family of five to buy whatever they want

Chapter Five | Emmanuel Ojameruaye, in
Jideofor Adibe (Ed.)
The Politics and Economics of Removing Subsidies on Petroleum Products in Nigeria
London & Abuja, Adonis & Abbey Publishers

119

in place of the subsidy! However, the government cannot pay out N8, 000 to each Nigerian as we do not yet have a system to ensure that such payments get to all citizens.

e) Promotes inefficiency: Subsidies on locally-produced goods can artificially protect inefficient firms who need to restructure. Thus, subsidies tend delay much needed economic reforms. For example, the payment of subsidies to inefficient fertilizer companies or steel companies or local refineries may protect them and thereby prevent them from making necessary changes to become more efficient. While this does not apply to the subsidy paid to importers of fuel, it could apply to subsidies paid to local refineries.

f) Increase in demand: Subsidies increase demand for (and consumption) of subsidized products which could discourage the demand for (and consumption) of alternatives or substitutes. For example, the existence of subsidies on petroleum products has led to an exponential increase in the demand for (and consumption of) petroleum products. This has discouraged or reduced the demand for, and investment in, alternative energy (or substitutes) such as coal, solar, wind, biomass, etc. If subsidy is reduced or removed from petroleum products, it could spur investment in alternative energy resources, especially renewable or green energy, which will in turn increase the consumption of renewable energy and conserve crude oil which is non-renewable. This is why some advanced countries impose heavy taxes on petroleum products in order to encourage invest in, and consumption of, renewable energy such as solar, wind, geothermal and hydro. (Note: Renewable energy is energy which comes from natural resources such as sunlight (solar), wind, rain, tides, water (hydro) and geothermal heat which are renewable (naturally replenished). About 16% of global final energy consumption currently comes from renewables)

Chapter Five	Emmanuel Ojameruaye, in
	Jideofor Adibe (Ed.)
	The Politics and Economics of Removing Subsidies on Petroleum Products in Nigeria
	London & Abuja, Adonis & Abbey Publishers

g) Uneven playing field: Paying subsidy to importers of petroleum products while local refineries do not receive an equal amount of subsidy creates distortions in the petroleum products market and uneven playing field between the local refineries and foreign refineries represented by the importers. This makes local refineries to sell their products at artificially low prices, and the refineries are therefore unable to generate adequate revenues to maintain their plants and expand their production capacities. Thus, the local refineries are not able carry out required turn-around maintenance (TAM) of their plants as at when due, and this has resulted in frequent breakdowns and persistent underutilization of their installed capacities. In fact, over the past 30 years, the local refineries have operated at less than 40% of their combined installed capacity most of the time. If all the refineries are able to operate at 90% of their capacities, they would probably produce enough petroleum products for domestic consumption, thereby eliminating or reducing the importation of petroleum products. It is argued that if fuel subsidy is reduced and the prices of petroleum products are allowed to increase to reflect international market prices, the local refineries will be able to generate enough revenues not only to maintain their plants to ensure over 90% capacity utilization but also to invest in the expansion of their production capacity and produce for exports. Thus, it can be argued that government should pay commensurate subsidy to the local refineries similar to what it pays to suppliers of imported petroleum products.

h) Private investment: Paying subsidy to importers of petroleum products while local refineries do not receive commensurate subsidy discourages private investors in the downstream sector because the low price of petroleum products makes it unprofitable for private investors to establish new refineries at huge cost only to sell their refined at low prices that cannot

Chapter Five | Emmanuel Ojameruaye, in
Jideofor Adibe (Ed.)
The Politics and Economics of Removing Subsidies on Petroleum Products in Nigeria
London & Abuja, Adonis & Abbey Publishers

guarantee adequate returns on investment. If the private refineries are to buy crude oil from NNPC or other upstream oil companies at international (spot) prices and refine the crude locally into petroleum products they cannot afford to sell the products at the "controlled" price because they will not break even or generate adequate income for maintenance and dividends. It is not surprising therefore that of the 26 companies that have been granted licenses to establish and operate local refineries in Nigeria since 2002, only one has recently managed to establish a small (mini-diesel) refinery (referred to as a topping plant) in Rivers State. All the other companies are still "waiting and watching" and some of them have become importers of fuel in order to benefit from the more lucrative subsidy windfall. Eighteen licenses were granted to private investors in 2002 to establish private refineries in the country. In 2004, another eight licenses were granted, bringing the total to 26. The construction of the mini-diesel refinery, owned by the Niger Delta Petroleum Resources (NDPR), started in January 2010 and was completed in December 2010. It is currently undergoing test-run. It can only refine 1,000 barrels a day of crude oil and produce 120,000 liters of diesel a day using crude oil from the nearby Ogbelle Flowstation (*ThisDay*, November 18, 2011)

i) Illegal activities: Fuel subsidy in Nigeria has also encouraged illegal activities such as smuggling of petroleum products out of the country to neighboring countries, adulteration of petroleum products, round-tripping of petroleum products, emergence of illegal small/cottage refineries in the creeks in the Niger Delta region and poor quality of petroleum products which damage of engines/machines and increase environmental pollution.

j) Shortages: Fuel subsidy has led to sporadic shortages of petroleum products as evidenced by frequent long queues at filling stations, signs of "no fuel" at filling stations, selling of

Chapter Five | Emmanuel Ojameruaye, in
Jideofor Adibe (Ed.)
The Politics and Economics of Removing Subsidies on Petroleum Products in Nigeria
London & Abuja, Adonis & Abbey Publishers

petroleum products at inflated prices along the highways and street corners. This happens when government is unable to provide subsidy to cover the importation of adequate volume of petroleum products to supplement the quantity produced by local refineries. Frequent shortages of petroleum products lead to increases in the cost of transportation and cost of production as well as interruptions in production and frustration of motorists and other users of petroleum products. For instance, there is currently an acute shortage of kerosene in the Lagos area, and a liter of kerosene is selling for N150 per liter outside the filling stations which have none to sell!

Global Perspectives of Petroleum Products Pricing and Subsidy

Nigeria is not the only country grappling with the issue of fuel subsidies and appropriate pricing of petroleum products. In fact, the pricing of petroleum products and fuel subsidies have become a major global issue because of the phenomenal increase in global fuel subsidies and the need to curb the demand for fossil fuels and reduce greenhouse gas (carbon dioxide) emissions. Global pre-tax fuel subsidy increased from about $60 billion in 2003 to about $300 billion in 2008 and then $409 billion in 2010. It may reach $660 billion in 2020 or 0.7% of global GDP (IEA, 2011). If we assume an optimal tax of $0.3 per liter[15], the corresponding global tax-inclusive subsidy is estimated to have increased from $410 billion in 2003 to $1,000 billion in mid-2008 and $520 billion in mid-2009, representing 0.9% of global

---

[15] In view of revenue and environmental considerations, it is necessary to factor "optimal taxation" in estimating the "true" subsidy. For a country, a fuel tax that is less than the "optimal tax" level generates a "tax subsidy". Energy economists estimate the magnitude of tax subsidies in each country and add it to the pre-tax subsidy to get the tax-inclusive subsidy. So, even though a country may not have pre-tax subsidy, it could have a tax-inclusive subsidy if its fuel tax is less than the optimal level, which varies from country to country, but estimated at between $0.2 and $0.3 per litre for some countries (Coady, 2010)

Chapter Five | Emmanuel Ojameruaye, in
Jideofor Adibe (Ed.)
The Politics and Economics of Removing Subsidies on Petroleum Products in Nigeria
London & Abuja, Adonis & Abbey Publishers

GDP (Coady, 2010). The "emerging" economies accounted for about 70% of pre-tax global subsidy in 2008 while the "developing" countries accounted for about the remainder 30% while there were no pre-tax subsidies in the advanced countries. On the other hand, emerging economies accounted for about 64% of tax-inclusive global subsidy in 2008 while developing countries accounted for about 23% and advanced countries for 13% (see table below).

Table 1 : Total World Fuel Subsidies			
	End-2003	Mid 2008	Mid 2009
	In nominal billions US dollars ($)		
Pre-tax subsidy	57	519	136
Tax-inclusive subsidy (at tax threshold of US0.3 per litre)	406	998	524
Tax-inclusive subsidy (at tax threshold of US0.4per litre)	579	1206	721
	In percent of total subsidy (%)		
Pre-tax Subsidy:			
Advanced Countries	0.2	0	0
Emerging Economies	66	70.3	60.5
Developing Countries	33.8	29.7	39.5
Tax-inclusive subsidy (at tax threshold of US0.3 per litre)			
Advanced Countries	36.7	12.7	26.5
Emerging Economies	45.6	63.9	52.4
Developing Countries	17.7	23.4	21.1

Source: Coady, *et al* (2010)

In general oil exporting countries tend to subsidize petroleum products while the member countries of the EU and other advanced countries tend to impose high taxes. The verdict from most energy economists is that all countries of the world should phase out pre-tax subsidies as soon as possible while efforts should be made to reduce tax-inclusive subsidies in the long-run. For instance, according to International Energy Association (IEA) estimates, "reducing (pre-tax) subsidies by one-half could reduce greenhouse gas emissions by nearly 5 % by 2050. Reducing tax-inclusive subsidies by one-half

Chapter Five | Emmanuel Ojameruaye, in
Jideofor Adibe (Ed.)
The Politics and Economics of Removing Subsidies on Petroleum Products in Nigeria
London & Abuja, Adonis & Abbey Publishers

124

would result in larger emissions reduction of 14-17 percent by 2050"
(Coady, 2010:12). Furthermore, the IEA notes that "subsidies are an
extremely inefficient means of assisting the poor: only 8% of the $409
billion spent on fossil fuel subsidies in 2010 went to the poorest 20%
of the population". The IEA has also estimated that the phasing out of
fuel subsidies by 2020 would a) slash growth in energy demand by
4.1%; b) reduce growth in oil demand by 3.7 mb/d; and c) cut growth
in Carbon Dioxide emissions by 1.7Gt (IEA, 2011). It is obvious from
the above that there is a wide variation in the domestic (retail) prices
of petroleum products in the different countries of the world
reflecting the different approaches to the pricing of petroleum
products in these countries and the different rates of fuel subsidy and
fuel taxation. For example, the table below shows the "subsidy
burden" varies considerably even among some the highly
"subsidizing countries" shown in the table below.

Table 2: Fossil Fuel Subsidies in Selected Countries in 2010

Country	Amount of subsidy in 2010 (US$ billion)	Average subsidy rate* (%)	Subsidy per person ($)	Total Subsidy as a Share of GDP (%)
Nigeria	2.44	28.3	18.4	1.3
Angola	0.94	31.5	59.1	1.3
Libya	3.17	71	665	5.7
Algeria	8.46	59.8	298.4	6.6
Saudi Arabia	30.57	75.8	1,586.60	9.8
Iran	40.92	84.6	1,093.10	22.6
Mexico	9.34	12.5	83.8	0.9
Venezuela	15.7	75.3	689.2	6.9
India	16.2	13.5	18.2	1.4
China	7.77	3.8	15.9	0.4
Iraq	8.87	56.7	357.3	13.8
Egypt	14.07	55.6	250.1	9.3
Total	**158.45**			

* Average subsidy rate is the amount of subsidy paid on a liter expressed as a
percentage of the supply cost. For example is the supply cost of gasoline is 80 US

Emmanuel Ojameruaye, in
Jideofor Adibe (Ed.)
The Politics and Economics of Removing Subsidies on Petroleum Products in Nigeria
London & Abuja, Adonis & Abbey Publishers

cents ($0.9) per liter and the retail price is 60 cents per liter, it means the subsidy amount is 20 cents per liter which is 25% of the supply cost (i.e. 20/80 x 100%). Subsidy per person is total subsidy divided by the population, i.e. per capita subsidy. *Source:* Coady, et al (2010)

The German Agency for International Development (*Deutsche Gesellschaft für Internationale Zusammenarbeit* or GIZ) provides a comprehensive country-by-country analysis of gasoline and diesel retail prices through its publications, *International Fuel Price*, based on periodic surveys and analyses of the retail prices of gasoline and diesel in 174 countries in the world. United States. It its most recent editions of the publication, the GIZ has grouped the countries surveyed into four categories based on the level of subsidies and taxation on gasoline and diesel (GIZ, 2012) as follows:

a) Category A - Very High Fuel Subsidies Countries: These are countries where the retail price of fuel (gasoline and diesel) is below the price of crude oil on the world market. In the November 2008 edition, these were countries where the retail price of gasoline (and diesel) was between 1 and 29 US cents per litre, i.e. below the world market crude oil price of 30 cents per litre as of November 2008. In the November 2010 edition, when the price of crude oil was 51 cents per litre, the range was 1 to 50 cents per litre. Among the country within this category of "High Subsidies" in the November 2010 were large oil exporting countries such as Venezuela, Saudi Arabia, Kuwait, Libya and Algeria.

b) Category B- Fuel Subsidies Countries: These are countries where the retail price of fuel is above the price of crude oil on the work market and below the average retail price of fuel in the United States. The retail fuel price in the US is used as a benchmark because fuel prices of the United States are

> ...average cost-covering retail prices including industry margin, VAT and 10 US cents for Federal and State road funs. This fuel

Chapter Five | Emmanuel Ojameruaye, in
Jideofor Adibe (Ed.)
The Politics and Economics of Removing Subsidies on Petroleum Products in Nigeria
London & Abuja, Adonis & Abbey Publishers

price may be considered as the international minimum benchmark for a non-subsidized road transport policy (GIZ, 2012).

In the November 2008 edition, these were countries where the retail price of gasoline (and/or diesel) was between 30 and 55 US cents per liter, i.e. above the price of crude oil on the world market but below the average retail price of gasoline (and/or diesel) in the United States (then 56 cents per liter). In the November 2010 edition when the average retail price of gasoline was 76 cents per liter (and 84 cents per liter for diesel), the range was 51 to 76 cents for gasoline (and 51 cents to 84 cents for diesel). Among the countries in this category (Subsidies) in the November 2010 edition were many other oil exporting countries like Malaysia, Indonesia, Mexico, Angola and Nigeria, as well as few non-oil exporting countries.

c)   Category C - Fuel Taxation Countries: These are countries where the retail price of fuel is above the price level is the U.S. and below the price of fuel of the cheapest European Union (EU) country. The country in the EU with the cheapest fuel is used as a benchmark because fuel prices in the EU countries include VAT, fuel taxes, as well as other country-specific duties and taxes. In the November 2008 edition, the retail fuel price in Spain was used as a benchmark because fuel prices in Spain were lowest (at 122 cents per liter for gasoline) in the EU-15 countries. However, for the November, 2010 edition, the retail prices of gasoline and diesel in Luxemburg and Romania, respectively, were used as benchmarks because they were lowest within the EU-27 countries (at 146 cents per liter for gasoline in Luxembourg and 136 cents per liter for diesel in Romania). Thus, the price range for this category was 76 to 146 cents for gasoline and 84 to 136 cents for diesel. The countries in this category (Taxation) in the November 2010 edition included Pakistan, India (diesel only, Australia, Argentina,

Chapter Five | Emmanuel Ojameruaye, in
Jideofor Adibe (Ed.)
The Politics and Economics of Removing Subsidies on Petroleum Products in Nigeria
London & Abuja, Adonis & Abbey Publishers

127

China PR, South Africa and all Nigeria neighboring countries
– Benin, Cameroon, Niger and Chad.

d) Category D - High Fuel Taxation Countries: These are
   countries where the retail price of fuel is above the cheapest
   price in the EU. In the November 2008 edition, the benchmark
   was Spain (at 122 cents) but for November 2010 the
   benchmark price was 146 cents for gasoline (in Luxemburg)
   and 136 cents for diesel (in Romania). The countries in this
   category (High Taxation) in the November 2010 edition
   included all the EU countries, Japan, Israel, Turkey, Uruguay,
   Malawi, Rwanda and Zambia. Table 3 and figure 1 below
   show the retail prices of gasoline and diesel for selected
   countries and the categories under which they fall in the
   November 2010 edition of the GIZ's *International Fuel Prices*
   publication.

Table 3: Comparison of Retail Fuel Prices in Selected Countries, Nov. 2010

Continent/Countries	Diesel Price (US Cents per litre)	Gasoline Price (US Cents per litre)	Average Fuel Price	Category
AFRICA				
Libya	13	17	15	A = High Subsidies
Algeria	32	19	25.5	A = High Subsidies
Angola	43	65	54	A/B
Nigeria	77	44	60.5	B/A
Ghana	83	82	82.5	B = Subsidies
Cameroon	121	104	112.5	C =Taxation
Benin	110	120	115	C =Taxation
Niger	116	107	111.5	C =Taxation
South Africa	114	119	116.5	C =Taxation
Zambia	152	166	159	D=High Taxation
Rwanda	162	163	162.5	D=High Taxation
ASIA/PACIFIC				
Iran	1.6	9.7	5.65	A = High Subsidies
Saudi Arabia	6.7	16	11.35	A = High Subsidies
Kuwait	21	23	22	A = High Subsidies
Indonesia	51	51	51	A/B
Malaysia	56	59	57.5	B = Subsidies
India	82	115	98.5	B/C
Pakistan	92	86	89	C =Taxation

Chapter Five	Emmanuel Ojameruaye, in
	Jideofor Adibe (Ed.)
	The Politics and Economics of Removing Subsidies on Petroleum Products in Nigeria
	London & Abuja, Adonis & Abbey Publishers

Australia	123	127	125	C =Taxation
Japan	137	160	148.5	D=High Taxation
AMERICAS				
Venezuela	1.1	2.3	1.7	A = High Subsidies
Ecuador	28	53	40.5	A/B
Bolivia	54	70	62	B = Subsidies
Mexico	74	81	77.5	B = Subsidies
USA	84	76	80	B/C
Argentina	105	96	100.5	C =Taxation
Uruguay	144	149	146.5	D=High Taxation
EUROPE				
Romania	136	156	146	C/D
Luxembourg	146	146	146	D/C
Germany	168	190	179	D=High Taxation
France	172	198	185	D=High Taxation
United Kingdom	198	192	195	D=High Taxation
Norway	201	212	206.5	D=High Taxation

Source: Compiled from GIZ (2012)

Chapter Five	Emmanuel Ojameruaye, in
	Jideofor Adibe (Ed.)
	The Politics and Economics of Removing Subsidies on Petroleum Products in Nigeria
	London & Abuja, Adonis & Abbey Publishers

Figure 2: Retail Prices of Gasoline and Diesel in Selected Countries, Nov. 2010

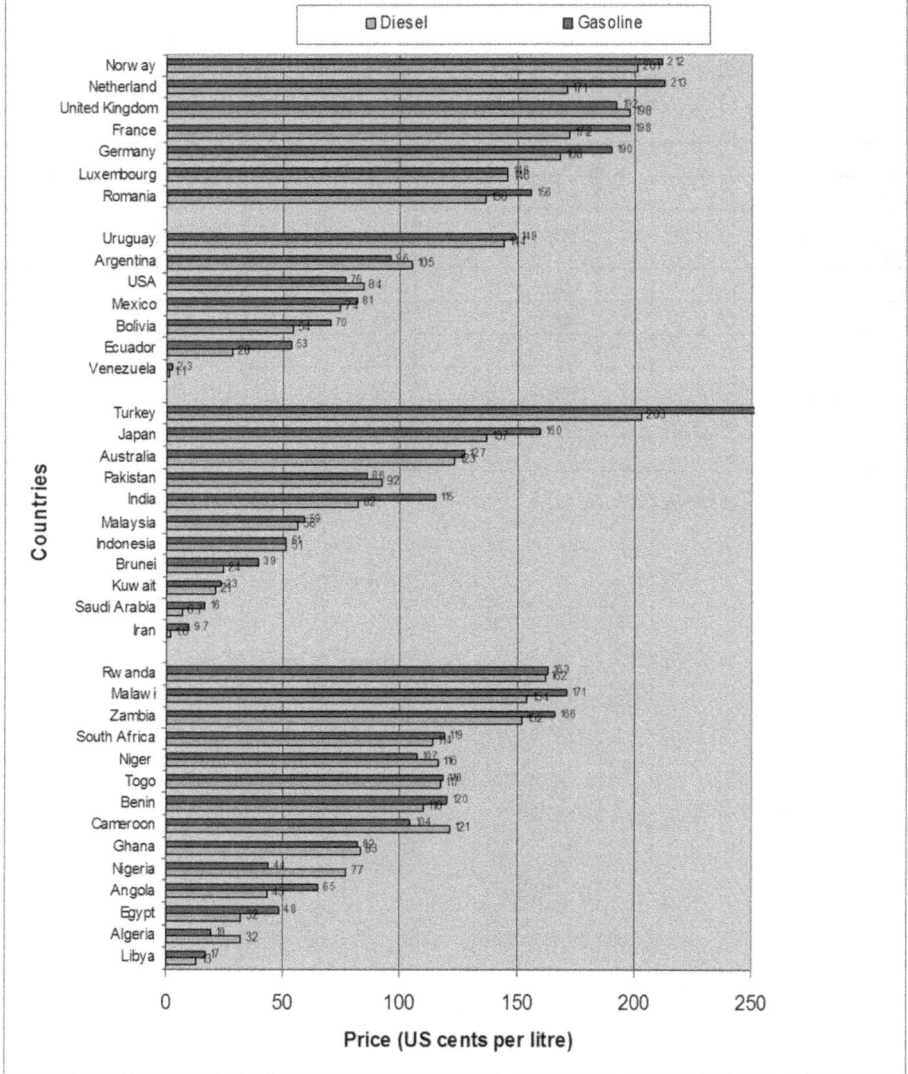

□ Diesel   ■ Gasoline

Countries (Norway, Netherland, United Kingdom, France, Germany, Luxembourg, Romania, Uruguay, Argentina, USA, Mexico, Bolivia, Ecuador, Venezuela, Turkey, Japan, Australia, Pakistan, India, Malaysia, Indonesia, Brunei, Kuwait, Saudi Arabia, Iran, Rwanda, Malawi, Zambia, South Africa, Niger, Togo, Benin, Cameroon, Ghana, Nigeria, Angola, Egypt, Algeria, Libya)

Price (US cents per litre) — 0, 50, 100, 150, 200, 250

Source: Charted with data from GIZ (2012)

GIZ and most economists believe that countries in the first two categories (high subsidy and subsidy) must transition to the other two

Chapter Five | Emmanuel Ojameruaye, in
Jideofor Adibe (Ed.)
The Politics and Economics of Removing Subsidies on Petroleum Products in Nigeria
London & Abuja, Adonis & Abbey Publishers

130

categories (taxation and high taxation) over time. To achieve this transition, GIZ has identified four basic principles that should inform the design of national fuel taxation policies and a rational fuel pricing scheme: The principles are

i)   Fuel prices should cover production and distribution costs. If a country implements this principle, its fuel prices should be very close to the upper limit of category B.

ii) Fuel taxes should help to finance the transport sector, according to the user pays principle. If this is implemented in a country, then fuel prices will almost equate those in the United States or a little above, i.e. at the lower limits of category C prices.

iii) Fuel taxes should help internalize external costs and incentivize energy efficiency in the transport sector. If this is implemented in a country, then fuel prices will be any from about the middle to the upper limit of category C prices.

iv) Fuel prices should contribute to the general revenues of government, e.g. normal VAT should be imposed on petroleum products. If this is implemented in a country, fuel prices will fall within the category D price range.

However the transition from category A to D may take a very long time because countries are using different pricing mechanisms to adjust fuel prices. GIZ has also identified three basic forms or mechanisms often used by countries to adjust fuel prices. These are:

a) Ad hoc pricing/regulation: In countries with ad hoc fuel pricing mechanism, fuel prices are set or adjusted unsystematically at irregular intervals or prices remain unchanged over several years even when crude oil prices and supply cost of petroleum products are changing frequently. In these countries, fuel pricing is highly political and the decision to adjust fuel prices is made by the government based largely on political and social

Chapter Five | Emmanuel Ojameruaye, in
Jideofor Adibe (Ed.)
The Politics and Economics of Removing Subsidies on Petroleum Products in Nigeria
London & Abuja, Adonis & Abbey Publishers

131

considerations. Most countries with this pricing mechanism end up subsidizing fuel. Countries with this mechanism include Saudi Arabia, Bolivia, Qatar and Nigeria.

b) Regular/active price adjustment/regulation: Under this mechanism, fuel prices are regulated but are reviewed and adjusted regularly based on pre-determined criteria and/or formulae. The adjustments can be made weekly or monthly or quarterly based on changes in the underlying price determinants. In most cases, the components of the price (cost, margins, tax, etc) and other factors underlying price changes as well as the adjustment intervals are set by statute (law). An independent institution may also be assigned the responsibility of monitoring the prevailing regulations and pricing formula. The decision to adjust prices is usually depoliticized and price adjustments are usually transparent, verifiable and understandable by the public. Examples of countries with this mechanism include South Africa, China and Vietnam.

c) Liberalized markets (i.e. passive or no regulation: Under this mechanism, the role of the government is simply to set taxes and excise duties on fuel and to monitor and ensure fuel quality and measurement standards. In other words, regulation is limited to setting the level of taxes and framework conditions such as quality of fuel and measurement standards and inspections. The retails prices of fuel are usually determined by the free market forces (demand and supply) and by the petroleum products companies. This retail fuel prices tend to fluctuate frequently as underlying demand and supply factors change, sometimes daily, weekly or monthly. Fuel prices are almost entirely depoliticized. Examples of countries with liberalized fuel markets are the EU countries, USA, Japan, Kenya, Uganda and the Philippines.

GIZ has recommended that for countries with ad hoc pricing, fuel pricing reform efforts should be based on a long-term (5 to 10 years) perspective beginning with a comprehensive transparency campaign. Subsidies could be maintained initially with regular

Chapter Five | Emmanuel Ojameruaye, in
Jideofor Adibe (Ed.)
The Politics and Economics of Removing Subsidies on Petroleum Products in Nigeria
London & Abuja, Adonis & Abbey Publishers

(monthly) price adjustments with the ultimate goal of phasing out subsidies by a given date and eventual introduction of fuel taxes with earmarks for transportation projects and social safety nets. On the other hand, countries with regular price reviews are encouraged to continue the regular price adjustments based on changes in input parameters as well as improve transparency and outreach to the public. Countries with passive or no regulation are encouraged to apply tools that increase transparency and limit daily fluctuations as well as potential profiteering by retailers, e.g. by publishing indicative maximum prices and costs of input products and margins. For high subsidies and subsidies countries,

> "Fuel subsidies are not only a burden on the economy but also encourage wasteful fuel use. Normalized petrol prices, in line with world market prices, would clearly reduce the impact of subsidies on the economy and foster an energy efficient transport system…A transparent fuel pricing scheme will prepare both the consumer and oil marketers for the frequent movement in market prices"(GIZ, 2012)

Coady *et al* (2010) have proposed some options for subsidy reform for Category A and B countries with ad hoc pricing mechanism. The reform must begin with the understanding that

> "There are gains from targeting transfers, as most of the benefits from universal petroleum subsidies accrue to high-income households. The benefits of gasoline subsidies are the most regressively distributed, with over 80 per cent of total benefits accruing to the richest 40 per cent of households…However, the elimination of even badly targeted subsidies can have an adverse impact on poor households, requiring the implementation of measures to protect the poor"(Coady *et al*, 2012).

Subsidy reform however faces several obstacles, including the following (Coady *et al* (2010) :

• Weak capacity to target mitigating measures to the poor;
• Lack of transparency in reporting subsidies;
• Opposition by vested interests;

Chapter Five | Emmanuel Ojameruaye, in
Jideofor Adibe (Ed.)
The Politics and Economics of Removing Subsidies on Petroleum Products in Nigeria
London & Abuja, Adonis & Abbey Publishers

133

- Cross-border spillover effects; and
- Reforming the (ad hoc) price-setting mechanisms.

Based on their study of petroleum products prices and subsidies, Coady *et al* (2010) reached the following key conclusions:

- Government direct control of domestic petroleum prices has often been an obstacle to subsidy reform.
- The first-best solution is to liberalize petroleum prices.
- If markets are imperfect or if governments are concerned about excessive price volatility they can implement an automatic pricing mechanism that adjusts prices regularly in light of changes in international prices.

Conclusion and Recommendations for Nigeria

The decision to eliminate or reduce or retain the petroleum products subsidy in Nigeria is not going to be an easy one. It is one that needs to be approached with caution after all the factors outlined in this chapter have been taken into consideration. The following are some conclusions and recommendations for dealing with the petroleum products pricing and subsidy issue in Nigeria:

a) The debate on the removal or elimination of fuel subsidy in Nigeria has tended to be more political than economic. The proponents and opponents of fuel subsidy have tended to adopt a "war" approach in their arguments and have used very little theory or established knowledge, data and evidence to justify their positions.
b) The testimonies and presentations of government departments (NNPC, PPPRA, FMF, FMPR) at the hearings in the National Assembly Committees, which looked into the issue as well as some of the utterances and claims by the Presidency, have left

Chapter Five | Emmanuel Ojameruaye, in
Jideofor Adibe (Ed.)
The Politics and Economics of Removing Subsidies on Petroleum Products in Nigeria
London & Abuja, Adonis & Abbey Publishers

134

much to be desired and have also painted a picture of insincerity and poor grasp of the issue.

c) The hearings at the National Assembly and the Report of the Ad-hoc Committee of the House of Representative clearly show that the management of fuel subsidy in Nigeria has been very poor, ineffective, and inefficient and marked by corrupt practices and poor data management. Therefore, an independent audit of the fuel subsidy system should be undertaken immediately by a reliable and reputable audit company, preferably a foreign one. The audit should include a re-verification of the amounts paid to importers and local refineries, evidence of quantity of petroleum products imported and supplied, cost of producing petroleum products by local refineries, the crude for product swaps, etc. There is also an urgent need to conduct an independent audit of NNPC and all its subsidiaries with focus on its revenues and expenditures as there are strong indications inefficiency and profligacy within that organization for which the Nigerian consumers are paying.

d) Removing the existing fuel subsidy under the current situation which is tantamount to increasing the pump price of petroleum products will amount to a transfer of the cost of government inefficiency and the burden official corruption to consumers of petroleum products and the poor will likely be the hardest hit because they have limited real income loss compensation mechanisms.

e) The National Assembly should reject any proposal to remove or reduce the fuel subsidy until the Presidency comes up with a credible plan to reduce corruption associated with fuel subsidy and improve the current system. The National Assembly should also demand the resignation of some government officials who have demonstrated gross incompetence in their testimonies during the fuel subsidy investigations. This will serve as a wake-up call for officials to

Chapter Five | Emmanuel Ojameruaye, in
Jideofor Adibe (Ed.)
The Politics and Economics of Removing Subsidies on Petroleum Products in Nigeria
London & Abuja, Adonis & Abbey Publishers

135

be on top of their jobs and work for the interest of the people rather than special interests.

f) The federal government must resist any temptation to remove or reduce the subsidy until it has secured approval of the National Assembly and other key stakeholders and organizations, including the labor unions. To this end, the federal government must first demonstrate sincerity on the issue and win the confidence of the people by putting in place a plan to improve the fuel subsidy system. The plan should include revitalizing the local refineries to operate at a minimum of 80% of their installed capacity within a period of one year, ensuring that at least three private refineries with at least a combined capacity of 150,000 barrels a day are established and operational within one year, gradual reduction in the volume of imported petroleum products, abolishing the current PPPRA pricing template and establishing a more competitive way of determining the level of subsidy. Once this is done, the federal government can come up with a plan to gradually phase out (reduce) the subsidy in a stepwise fashion over a period of two years.

g) The federal government should revoke all the import licenses granted to private importers of petroleum products with immediate effect and appoint the Products and Pipelines Marketing Company (PPMC), an NNPC's subsidiary, as the sole importer of petroleum products into the country. The activities of PPMC should be closely monitored by the National Assembly and other watch-dog groups. The PPMC should adopt proven techniques to determine the monthly demands for the various types of petroleum products, domestic production, and the required level of imports to bridge the gap between demand and domestic production. For instance, if the projected demand for petrol in March 2012 is 200,000 barrels a day (= 32 million liters a day) or 6.2 million barrels for that month (= 992 million liters), and if local

Chapter Five | Emmanuel Ojameruaye, in
Jideofor Adibe (Ed.)
The Politics and Economics of Removing Subsidies on Petroleum Products in Nigeria
London & Abuja, Adonis & Abbey Publishers

136

refineries will produce 80,000 barrels a day or 2.48 million barrels for that month, then PPMC can place order for 120,000 barrels a day or 3.72 million barrels (or 591 million liters) for that month. If the projected landing cost is $100 per barrel (about N100 per liter), then PPMC will need about $380 million to import petrol for that month.

h) The federal government should direct NNPC to stop the "crude for products swap arrangements" - the practice of swapping crude oil for refined petroleum products from foreign refineries- with immediate effect. Under these "cashless" deals, NNPC provides crude oil to refineries located outside the country to refine and the refineries then ship the refined products to Nigeria. It is like modern-day trade by barter. It is inefficient and prone to corruption and cheating. It is one of the sources of "round tripping" of crude and petroleum products. In fact, during the hearing at the Senate Committee, the NNPC Group Managing Director (GMD) could not account for 65,000 barrels a day that was allegedly swapped. At the Senate hearing the GMD revealed that under the Crude for Products Swap arrangement, out of the 445,000 barrels a day of crude oil allocated to NNPC to refine for domestic consumption, it allocates 170 barrels a day to local refineries (Warri -80,000, Port Harcourt – 90,000) and 210,000 barrels to refineries located outside the country (Duke Oil -90,000 barrels per day, Transfigura UK – 60,000 barrels a day; and Societe Ivorien Refineries (SIR). an Ivorian refinery - 60,000 barrels per day) which left a balance of 65, 000 barrels a day unaccounted for. According to the GMD,

> When you take the crude, you sell on the market at official price and use the money to import products but in a swap arrangement, there's flexibility in the products that you get. In a swap arrangement, the transaction is cashless. (*Daily Sun*, December 13, 2011).

Chapter Five — Emmanuel Ojameruaye, in
Jideofor Adibe (Ed.)
The Politics and Economics of Removing Subsidies on Petroleum Products in Nigeria
London & Abuja, Adonis & Abbey Publishers

However, a day after that event, the NNPC's spokesman, the late Dr. Levi Ajuonuma denied that the corporation could not account for 65,000 barrels a day. Notwithstanding this denial which also raises some issues, the swap arrangement is clearly not an efficient one. Of the 445,000 barrel a day of crude oil allocated to NNPC, any quantity that cannot be refined locally should be exported, that is, it should be sold at the international market rate, and the proceeds (cash) should be used to fund the subsidy on petroleum products imported by PPMC. For instance, of the 445,000 barrels a day allocated to NNPC, if the local refineries can only refine 170,000 barrels a day, NNPC can sell the balance 275,000 barrels a day at the spot market rate, say $95 per barrel, to make about $26 million a day or about $800 million a month which is more than enough to pay for over 250,000 barrels a day of imported products by PPMC, and which may be adequate to bridge the gap between demand and domestic production. In fact, when this is done, the federal government may not have to pay subsidy to PPMC directly to import petroleum products or it may pay a minimal amount if demand exceeds the quantity of petroleum products that can be produced from 445,000 barrels per day of crude oil allocated to NNPC.

i) Given that Nigeria produces about 2,400,000 barrels of crude oil a day, of which at least 1,200,000 barrels a day is allocated directly to the federal government (NNPC) as its share in the Joint Venture (JV) upstream oil companies (Shell, Mobil, Chevron, Agip, Elf, etc.) that produce the crude oil, the federal government can afford to allocate 500,000 barrels a day to NNPC to deliver to local refineries to refine locally, and then give the quantity that cannot be refined locally to PPMC to sell in the spot market and use the proceeds to import petroleum products to bridge the gap between demand and domestic production. If consumers pay 50% of the supply cost of the 500,000 barrels a day, the federal government will still have

Chapter Five | Emmanuel Ojameruaye, in
Jideofor Adibe (Ed.)
The Politics and Economics of Removing Subsidies on Petroleum Products in Nigeria
London & Abuja, Adonis & Abbey Publishers

138

over 700,000 million barrels a day to sell (and make at least $66.5 million a day or $2 billion a month or about $25 billion a year) in addition to petroleum profit tax (85%) it will collect from the 1.2 million barrels a day of crude oil that the JV partners and other oil companies operating under production sharing contracts (PSC) will sell and which can bring in at least another $15 billion in oil revenue net of JV cash calls.

j) The federal government should revoke the licenses to operate local refineries granted to companies that are yet to start construction activities. We do not need additional 26 local refineries in the country. The economies of scale dictate that we certainly do not need "mini-refineries" that can refine less than 20,000 barrels of crude a day. Given the fact that the current four NNPC-owned refineries have a combined installed capacity of 445,000 barrels of crude a day, and Nigeria's requirements (demand) for petroleum products can be meet by refining about 500,000 barrels of crude a day, we do not need more than five new medium to large scale private refineries with a combined capacity of 200,000 day (average of 40,000 barrels a day). Therefore, the federal government should grant new licenses to at most five companies that will be required to start construction activities within 12 months and start production with 24 months. If they fail to meet this condition, their licenses will expire without a renewal option.

k) Crude oil should be delivered to all local refineries (NNPC-owned and private) at international market (spot) prices and an agreed amount of subsidy or subvention that is evidence-based should be paid to the refineries by the federal government provided they are able to meet certain performance and cost benchmarks. Ultimately, the subsidy will be reduced gradually as the market becomes fully deregulated.

Chapter Five | Emmanuel Ojameruaye, in
Jideofor Adibe (Ed.)
The Politics and Economics of Removing Subsidies on Petroleum Products in Nigeria
London & Abuja, Adonis & Abbey Publishers

l) The variation in the prices of the different petroleum products in Nigeria should be minimal, and similar to the variation in the spot market prices, in order to avoid sharp practices such as products adulteration and eliminate or minimize cross-subsidization.

m) The subsidy on petroleum products should be reduced gradually in a stepwise fashion to zero over a period of two years. At the end of the two year period, a second-best pricing approach based on the Ramsey Pricing Principle should be adopted (Ojameruaye, 2013). Under this approach, the prices of petroleum products will be completely deregulated and the refineries will supply petroleum products to distributors at their supply costs plus their margins. Distributors will also supply products to retailers at different prices depending on their cost, location of retailers and their margins. Similarly, retailers will sell products at different prices in different cities or locations and within the same location. However, because of competition, the price variations will be small and inefficient refineries, distributors and retailers will be forced to close or become more efficient

n) Finally, the federal government must live up to its responsibility of securing the country's borders and prevent the smuggling of petroleum products out of the country or into the country. To this end, special border (and maritime) guards or patrol units should be established to patrol our borders and the coasts to prevent smuggling of petroleum products. These units should be well armed and equipped with modern telecommunication equipment, fast vehicles, boats, helicopters, and drones to intercept smugglers and illegal bunkerers. They should also be closely monitored to prevent their officers from being involved in corrupt practices and from aiding and abetting smuggling.

Chapter Five   Emmanuel Ojameruaye, in
                Jideofor Adibe (Ed.)
                The Politics and Economics of Removing Subsidies on Petroleum Products in Nigeria
                London & Abuja, Adonis & Abbey Publishers

References

Adenikinju, A (2009), 'Energy pricing and subsidy reform in Nigeria' (a presentation at the *Global Forum on Trade and Climate Change*, OECD Centre, 9-10 June. Available at:http://www.oecd.org/dataoe cd/58/61/42987402.pdf (Accessed: 5 April, 2013)

CBN (2012), 'Half Year Report for Jan- June' (Abuja, CBN). Available at: http://www.cenbank.org/   (Accessed: 20 April, 2013)

Coady, D. *et al* (2010), 'Petroleum product subsidies: Costly, inequitable and rising' (IMF, Fiscal Affairs Department), February 25.

GIZ (2012), *International fuel prices 2010/11*, 7th edition Available at: http://www.giz.de/Themen/en/29957.htm (Accessed: 25 April, 2013)

GIZ (2010), *International fuel prices 2008/9*. 6th edition Available at: http://www.giz.de/Themen/en/29957.htm (Accessed: 25 April, 2013)

IEA (2006), *Petroleum product pricing in India: Where have all the subsidies gone?* (OECD/ International Energy Agency).

IEA (2011), 'Analysis of fossil-fuel subsidies: World Energy Outlook, 2011' (International Energy Agency).

Kojima, M (2009), 'Changes in end-users petroleum product prices: A comparison of 48 countries'. Extractive industries and development series, #2. Feb. (Washington, The World Bank).

Makwe, I.A. (2006), 'A critique of the Nigerian Petroleum Products Pricing Regulatory Agency (PPPRA) Pricing Template and Cost Recovery Analysis'. *Oil, Gas and Energy Law* (OGEL), Vol. 3, September.

Ojameruaye, E.O. (2011), 'The political economy of the removal of petroleum products subsidy in Nigeria. Parts I and II'. Available at:___http://chatafrik.com/articles/economy/item/312-the-political-economy-of-the-removal-of-petroleum-products-subsidy-in-nigeria-part-i-the-politics.html

Chapter Five | Emmanuel Ojameruaye, in
Jideofor Adibe (Ed.)
The Politics and Economics of Removing Subsidies on Petroleum Products in Nigeria
London & Abuja, Adonis & Abbey Publishers

Ojameruaye, E.O. (2012a), 'The Petroleum Subsidy Probe Report: Some Technical Concerns and Weaknesses". Available at: http://chatafrik.com/articles/nigerian-affairs/item/929-the-petroleum-subsidy-probe-report-some-technical-concerns-and-weaknesses.html

Ojameruaye, E.O. (2012b), "Removing Petroleum Products Subsidy in Nigeria, To be, or Not to be". Available at: http://chatafrik.com/art icles/nigerian-affairs/item/1304-removing-petroleum-products-subsidy-in-nigeria-to-be-or-not-to-be?.html

Ojameruaye, E.O. (2013), "A Second-Best Framework for Petroleum Products Pricing in Nigeria" (aper presented at the 2013 Conference of the Nigerian Economic Society, September 17-19).

OPEC (2011), *Annual Statistical Bulletin, 2010/11* (Vienna, OPEC).

Peterson, W.W.F. (2009), *A Billion Dollars a Day: The Economics and Politics of Agricultural Subsidies* (NJ, Wiley-Blackwell).

Victor, D. (2009), 'Untold Billions: Fossil-Fuel Subsidies, Their Impacts and the Path to Reform. The politics of fossil-fuel subsidies'. (Paper prepared for the *Global Subsidies Initiative* (GSI), The International Institute for Sustainable Development (IISD), Geneva, Switzerland). October .

World Trade Organization (2006), 'The economics of subsidies' in *2006 World Trade Report: Exploring the links between subsidies, trade and the WTO*, Available at: http://www.wto.org/english/res_e/boo ksp_e/anrep_e/world_trade_report06_e.pdf (Accessed: 15 April, 2013)

Chapter Five | Emmanuel Ojameruaye, in
Jideofor Adibe (Ed.)
The Politics and Economics of Removing Subsidies on Petroleum Products in Nigeria
London & Abuja, Adonis & Abbey Publishers

142

CHAPTER SIX

Inclusive Growth and the Contradiction of Petroleum Product De-subsidisation Strategy

Abdelrasaq Nal

Introduction

In his 2012 budget speech that was delivered to a Joint Session of the National Assembly on December 13, 2011, the President of Nigeria, Dr. Goodluck Jonathan, described the 12-month economic blueprint as a product of his administration's desire to accomplish a pattern of growth that is inclusive (Business News, 2012). Adopting the term 'fiscal consolidation, inclusive growth and job creation' as slogan, the President understandably demonstrated the seriousness of his regime's commitment to the cause of this growth trajectory.

Less than a month later, precisely on January 1, 2012, in what can be interpreted as the administration's first strategic move towards realising the goal of inclusiveness, a removal of all existing subsidies on petroleum products was announced (Vanguard, 2013). This announcement provoked widespread condemnations, demonstrations and strike actions by individuals, the national labour union and other 'right activists' to protest the action. The immediate, direct and inevitable consequences of all these to the economy were massive loss of man-hours, properties and even lives across the country.

While it may be safe to argue at this juncture, that the Nigerian leader's fascination with the goal of inclusiveness is consistent with recent emphasis in development literatures that for economic growth to be sustainable and poverty reducing, its process and benefits must be shared, possibly pro poor and generally broad based across large part of a country's economy, the same cannot be said of the adoption

Chapter Six | Abdelrasaq Nal, in
Jideofor Adibe (Ed.)
The Politics and Economics of Removing Subsidies on Petroleum Products in Nigeria
London & Abuja, Adonis & Abbey Publishers

of petroleum product de-subsidisation as a strategy for accomplishing its objective.

Quite contrarily, this chapter will demonstrate that the action taken by Jonathan's administration on the New Year's Day of 2012 to remove subsidy on Premium Motor Spirit (PMS) has all the trappings of a scheme that is designed to exclude rather than include the poor majority from economic activities. The consequence of such exclusion will undoubtedly echo in increased poverty level and worsened inequality gap in the country.

In the next section, a brief introduction to the concept of inclusive growth is provided. Thereafter in Section Three, its framework is applied to analyse some key issues in the petroleum (oil) subsidy debate before the final section (Section Four) presents the Chapter's concluding remark.

The Inclusive Growth Paradigm

It must be clarified from the outset that the issue brought to the fore by the subsidy removal on petroleum products is not neoclassical economics. As such, it is not about the merits and demerits of subsidy removal: neither about fiscal discipline that blocks leakages in government finances nor about freeing up resources for development projects. It is not even about eliminating so called distortions that, according to Nigerian government officials, have prevented the flow of private investments into the downstream end of the oil sector. If there is anything the development experience of the 1990s taught us in emphatic terms, it is that sustainable economic progress is not about rigid compliance with the 'winner takes all' mentality of the neoclassical market philosophy. In the new perspective, good economics and indeed, sustainable economic progress requires policy makers to show a great deal of sensitivity to many of the socio-political concerns of development.

Interestingly, despite its limitations, the inclusive growth paradigm to development analysis is reasonably grounded in some

Chapter Six | Abdelrasaq Nal, in
Jideofor Adibe (Ed.)
The Politics and Economics of Removing Subsidies on Petroleum Products in Nigeria
London & Abuja, Adonis & Abbey Publishers

logics of political economy of reforms that one wonders why a self-proclaimed crusader of its ideals has refused to be allowed to be guided by its principles. As a concept that emerged out of frustration with the failure of neoclassical model to satisfactorily resolve some fundamental problems of development, especially with regards to the issue of poverty and inequality, inclusiveness provides that growth performance should be evaluated not only by the yardstick of pace but also pattern. Pattern in this sense refers to when growth is associated with progressive distributional change that helps bring about reduction in poverty and inequality.

A desirable growth pattern is one which permits large percentage of a country's labour force to contribute to its process and benefit from its outcome. The reason being that such an outcome is capable of not only reducing the poverty level but also the inequality gap. Anything inferior is considered toxic as it carries the dangerous potentials of derailing the course through political channel or conflict (World Bank, 2009). It is this emphasis on distributional consequences that distinguishes inclusive growth from the traditional neoclassical framework and provides a basis for addressing some of the socio-political concerns of reforms.

Chapter Six | Abdelrasaq Nal, in
Jideofor Adibe (Ed.)
The Politics and Economics of Removing Subsidies on Petroleum Products in Nigeria
London & Abuja, Adonis & Abbey Publishers

Box 1: Concept of Inclusive Growth

Inclusive growth is a pattern of growth that increases the opportunity for a large proportion of a country's population to participate in its process and benefit from its outcome (World Bank, 2009).

While the preoccupation of traditional neoclassical framework is with growth in its pure form, the inclusive growth paradigm is concerned with growth with positive distributional consequence. Past experience shows that pure growth alone is not enough to bring about satisfactory reduction in poverty and inequality.

A growth pattern that leaves a large number of people in poverty or cannot bring them out of it stands the risk of becoming unsustainable.

Sustainability of growth is possible when a large percentage of people (labour force) is carried along in its generation and proceed sharing. In other words, for growth to be sustainable in the long run it must be generated by the combined activities of a large majority rather than those of a few minorities.

What this implies is that a pattern of growth that relies on say high skill competences which can only be supplied by a few highly trained workers will be inferior to a pattern that is driven by say low skill competences that are generally possessed by many.

If growth is driven by the few, its process can be derailed by the large number it leaves behind. If a large number of people cannot participate in productive activities needed to survive, they are more likely to court survival through participation in destructive activities like terrorism, vandalism, robbery, etc.

Although the pace of growth is also acknowledged to be an important determinant of poverty alleviation in the inclusive growth literature, its primacy over pattern that some scholars subscribe to does not enjoy widespread acceptance.

Chapter Six | Abdelrasaq Nal, in
Jideofor Adibe (Ed.)
The Politics and Economics of Removing Subsidies on Petroleum Products in Nigeria
London & Abuja, Adonis & Abbey Publishers

Inclusive Growth in the Context of the Petroleum Subsidy Debate

The inclusive growth framework begins with the assumption that, at any point in time, and in any economy, various forms of distortions exist. Dealing with these distortions in a welfare maximising manner requires a rethink of our approach to policy reform in a number of areas. Some of these include finding the right answers to such questions as:

1. *Which among policy alternatives do we choose?*

The diversity of distortions means that their restrictive impacts on inclusive growth are of different magnitudes. Some have greater impacts than others. In Nigeria for instance, oil subsidy related distortions (OS) do not have the same impact on growth as distortions created by excessive remuneration of political office holders. We refer to the latter as 'Welfare Subsidy for Politicians' (WSP). In the same vein, the magnitude of damages that corruption inflicts on an economy may be bigger or smaller than damages associated with rent-seeking policies. Given the ineffectiveness, and in some cases, infeasibility of a reform agenda that targets all distortions at the same time, policy measure calls for eliminating the most important or binding ones (Ianchovichina and Lundstrom, 2009). This could be just one constraint that poses the biggest obstacle to growth or a combination of two or more.

But what is the appropriate yardstick for determining a binding constraint? The answer depends on whose script you are reading. The World Bank which has been championing the cause of inclusiveness since 2008 believes that the same neoliberal ideal of pace of growth is the most powerful tool for addressing poverty and inequality problems associated with growth process. To the Bank, unqualified growth is both necessary and sufficient for delivering on the objectives of growth with positive distributional consequences:

Chapter Six | Abdelrasaq Nal, in
Jideofor Adibe (Ed.)
The Politics and Economics of Removing Subsidies on Petroleum Products in Nigeria
London & Abuja, Adonis & Abbey Publishers

policy makers who seek to accelerate growth in the incomes of poor people and thus reduce overall poverty levels would be well advised to implement policies that enable their countries to achieve a faster rate of overall growth. A successful pro-poor growth strategy would thus need to have, at its core, measures for sustained and rapid economic growth. These measures include macroeconomic stability, well-defined property rights, trade openness, a good investment climate, an attractive incentive framework, well-functioning factor markets, and broad access to infrastructure and education. (Besley and Cord 2007, p. 19).

If you believe this, chances are that you would believe anything and 'growth obstruction' impact of distortion would be your yardstick. On the other hand, another school of thought associated with the pro-poor campaign places distributional concern at the forefront[16]. If you identify with this, you would use 'distribution obstruction' as your yardstick.

At the risk of simplifying the complexity of the exercise we apply these insights to reflect on the Nigerian situation. Assume the choice facing authorities is between removing either OS or WSP mentioned above. The legitimacy of these examples derives from the fact that they represent two forms of distortions that featured prominently in the debate that ensued with subsidy removal.

With respect to 'distribution obstruction' OS benefits a very large proportion of population and its retention relative to its removal does not make the country worse-off in terms of poverty and inequality. It can even be argued that its removal would contribute to widening the inequality gap. Therefore, it does not constitute a constraint to equitable distribution of income. On the other hand, WSP benefits only a small fraction of Nigerians who belong in the high income class. It represents an important constraint because its presence not only widens the inequality gap but also makes the poor worse-off in absolute terms. Targeting it for removal will at least close the income

---

[16] For an overview of the concept of pro-poor growth see OECD, (2006) and for exposition to the debate see Kakwani, et. al., (2004), Ravallion, (2004) and Ravallion and Chen, (2003).

Abdelrasaq Nal, in
Jideofor Adibe (Ed.)
The Politics and Economics of Removing Subsidies on Petroleum Products in Nigeria
London & Abuja, Adonis & Abbey Publishers

gap. On the basis of this casual analysis we can conclude that WSP in relation to OS is the binding constraint.

With respect to 'growth obstruction', one way of addressing the issue is to examine the impact of both forms of subsidies on consumption habits of recipients. It is a common knowledge that Nigerian politicians are ostentatious in consumptions. They buy foreign goods, send their children to foreign schools, acquire mansions abroad and even rush to foreign hospitals to treat the commonest of headaches. Therefore, retaining WSP will only serve to entrench these extravagant lifestyles that do nothing but put pressures on the exchange rates and take Nigerian jobs abroad. It is important to emphasise this aspect of loss of Nigerian jobs because job creation is at the heart of the transformation agenda of Jonathan's administration. In the contrary, much of the spending associated with OS is likely to benefit the economy as the poor people that constitute majority of the beneficiaries basically patronise Nigerian goods and services. On this score again, WSP is relatively the binding constraint to economic growth.

Now, let us look at some of the myths that have been peddled with respect to growth obstruction impact of oil subsidy regime. Specifically, we are taking on the issues of price distortions and who benefits most from subsidy.

It is claimed that subsidy related distortions that keep price artificially low is responsible for lack of private investment in the downstream sector. First and foremost, it must be pointed out that in many developing countries with functional refineries the state plays a crucial role. Nigeria did this successfully in the past and there is no evidence that commercial viability is the reason behind the collapse of her refineries. Secondly, if energy subsidies were that bad, why are their incidences so prevalent everywhere including the developed countries?

Moving to the issue, while it might be difficult to fault the claim in logic it is tempting to dismiss it as baseless in reality. Up till 2011

Chapter Six

Abdelrasaq Nal, in
Jideofor Adibe (Ed.)
The Politics and Economics of Removing Subsidies on Petroleum Products in Nigeria
London & Abuja, Adonis & Abbey Publishers

149

Nigeria sold fuel at N65 per litre. It is not clear whether, in relation to what ought to be the true market price this figure is higher, lower or an accurate reflection. Accounting figures for the year 2010 provide some clues. According to the International Energy Agency (IEA), fossil fuel consumption subsidisation rate in Nigeria in 2010 was 28 per cent of full cost of supply (IEA, 2012). This means true price would have hovered around N90 per litre. At N65 that fuel was sold during the year, we can say that the price was truly subsidised. But the IEA figure relied on price-gap technique that uses international price as benchmark. This approach has been widely criticised for its insensitivity and overestimating true value of subsidies in oil producing nations. What ought to constitute appropriate benchmark for these countries is their internal cost of production that would, under normal circumstances, be lower than international price. The law of comparative advantage which argues that a resource endowed nation cannot be as inefficient as a non-endowed one in the production of good that uses its endowment intensively is all about this. Both Professor Sam Aluko and OPEC have openly endorsed the view just as the IEA has also recently conceded to it.

What this implies is that the 28 percent subsidisation figure for 2010 is truly an overestimation and true price of fuel in Nigeria for the reference period could not have been up to N90 per litre. If we utilise correct data, we cannot rule out the possibility that the price would even be below the N65 figure that a litre was sold for.

Perhaps the best way to dismiss the government's claim is to ask: why is it that despite the absence of subsidies in other sectors like manufacturing, private investments have remained desperately low? If the truth must be told, private investment issue in Nigerian refineries extends far beyond subsidy. It touches on wider fundamental problems of corruption and generally poor investment climate in the country.

Finally, we find it quite simplistic and unconvincing to argue, on the basis of direct consumption, that it is the rich that benefits most

Chapter Six | Abdelrasaq Nal, in
Jideofor Adibe (Ed.)
The Politics and Economics of Removing Subsidies on Petroleum Products in Nigeria
London & Abuja, Adonis & Abbey Publishers

from subsidy induced lower fuel prices. If sufficient account is taken of the indirect benefits that transmit through the channel of reduced prices of locally produced goods and services that the poor, and only the poor Nigerians, depend on for survival, this argument will lose its appeal.

2. How do we implement our choice?

One of the key lessons of our experience with the failed Washington Consensus reforms is that sequence and pace of policy implementation matter[17]. When confronted with the challenge of how best to implement two or more reforms, it is required of policy makers to have good understanding of the nature of interactions among them. Some could be complementary to one another, and could be implemented in parallel.

The Nigerian government's attempts to improve transparency through the Freedom of Information Bill and reduce corruption through the establishment of an anti-corruption agency called Economics and Financial Crime Commission (EFCC) are good examples of complementary policy reforms that can be implemented simultaneously. Others may exhibit dependent relationship in that success of one depends on having previously achieved reasonable success in the implementation of another. This latter type will require sequencing. For instance, it can be argued that a dependent relationship exists between corruption and fuel subsidy and it would be well advised to implement reforms targeting the two distortions in

---

[17] The term **Washington Consensus** describes an economic philosophy with heavy attachment to the tenets of the free market and the presumption of the State as a source of both inefficiency and corruption. It comprises a set of reform package that was widely advocated in the 1990s for the crisis-ridden developing countries by the Washington based institutions like the International Monetary Fund (IMF), World Bank and the US Treasury Department. These reform measures included such market-oriented policies as, deregulation, fiscal discipline, tax reform, privatization, financial liberalization, trade liberalization, openness to FDI, among others.

Chapter Six	Abdelrasaq Nal, in
	Jideofor Adibe (Ed.)
	The Politics and Economics of Removing Subsidies on Petroleum Products in Nigeria
	London & Abuja, Adonis & Abbey Publishers

151

appropriate sequential order. From revenue generation point of view, going for oil subsidy before corruption may be inefficient as corruption that is left behind may fight within to wipe out all accomplishments through embezzlement of proceeds of reform.

Put differently, a cattle ranch owner whose ranch is infested with wolves and suffered significant depletion of his livestock due to foraging activities of the wolves cannot ask to be entrusted with more cattle, while the wolves are still present. If he wants more cattle, he needs to get rid of the wolves first.

Furthermore, care must be taken when deciding on the pace of implementation. The big bang approach where a given reform is implemented fully in one fell swoop is surely, not favoured. Gradualism or phased execution is a virtue that even some native neoclassical economists subscribe to. It allows for learning adjustments and softens potential economic hardship on those who stand to lose out from the reform. This fact is not new. Very early in the development of modern economics, it was recognised and emphasised by no other person than the great philosopher, Adam Smith. A cold-turkey implementation of reform is, in his view, not only likely to bring about some transitory inconveniences but also permanent losses.

Nigerian policymakers need to take some clues from their counterparts in developed countries with respect to how they handle pressures of energy subsidies withdrawal. In 2009, the G20 asked its members to rationalise and phase out inefficient fossil fuel subsidies. While the commitments of some members after more than two years are still at the level of proposal, others have simply put forth various arguments to exclude their policies from reform. Even among those that appear serious, no one is set to follow a big bang implementation strategy.

In June 2010 when members revealed their subsidy withdrawal plan at the Toronto summit, they all spaced out their implementation programme to cover different timelines. For instance, Canada

Chapter Six | Abdelrasaq Nal, in
Jideofor Adibe (Ed.)
The Politics and Economics of Removing Subsidies on Petroleum Products in Nigeria
London & Abuja, Adonis & Abbey Publishers

proposes a 5 year removal programme that commences in 2011 and ends in 2015. Germany proposes to complete its own programme by 2018. On the part of Russia, her plan to rationalise and phase out fossil fuel subsidies is expected to be completed in 2030 (Lang, 2011).

3. Who should participate in decision making?

A final rule of inclusive engagement and probably the most important is the requirement that policy making process must be participatory and inclusive of the often excluded members. Within the context of the traditional approach, danger is ever present that policy choice would be suboptimal because the more powerful stakeholders may hijack the process and make choices that serve their own selfish interests. To forestall this, decisions on such matters as objective of reform, choice of policy and implementation strategies must be taken in conjunction with representatives of the non-traditional constituencies like civil societies, labour unions, NGOs, etc. This is required, among others, to ensure better alliance of decisions with the interest of majority. Aside from this, the strategy also enhances accountability and transparency of governance, as well as increases chance of successful implementation.

It is clear that the administration of President Jonathan was not guided by this wisdom in its decision to remove fuel subsidy. Or how else, can one explain the fact that opposition to the reform is largely a roll call of usual suspects: civil society; trade unions; professional bodies; right activists, etc? A government that has only the privileged members of its society like state governors, cabinet ministers and a few powerful individuals as supporters and advocates of its policy surely has much to learn from the teachings of political economy of reforms.

Chapter Six | Abdelrasaq Nal, in
Jideofor Adibe (Ed.)
The Politics and Economics of Removing Subsidies on Petroleum Products in Nigeria
London & Abuja, Adonis & Abbey Publishers

Conclusion

This Chapter has highlighted some glaring inconsistencies in the management of Nigerian economy. It is argued that contradictions exist in the professed economic framework of inclusive growth and the removal of petroleum subsidy that the Jonathan administration put into effect in January, 2012. Specifically, the framework of inclusive growth provides that reforms must not be carried out with the mindset of 'business as usual'. Whether it is in the choice of policy action or decision makers or further still, how decisions reached are implemented, inclusive paradigm provides specific guidelines on what to and what not to do. Unfortunately, as the chapter has shown, the decision to remove subsidy on petroleum as well as the manner of its implementation clearly violated these guidelines.

What this has exposed is a gap in professed policy intentions of government and actions usually taken to bring them into fruition. But can we say, in the present case, that this is due to lack of knowledge of the right thing to do? We are pretty sure that this is not the case. In the alternative, is it a case of knowing what to do but lacking the political will to do it? We cannot say that this is not the case.

Many leaders fail not because the ideas needed to succeed are not out there but because due to some misguided considerations or weak governance capacities they often fail to apply them. It is an issue that bothers on credibility of commitment. Averting future recurrence of the needless controversies, disruptions and loss of lives and properties that the subsidy quagmire generated will depend to a great extent on the sincerity of government to policy making and execution.

References

Besley, T. and Cord, L. J. (2007), *Delivering on the promise of pro-poor growth: insights and lessons from country experiences*, (Washington, D. C., and World Bank).

Chapter Six | Abdelrasaq Nal, in
Jideofor Adibe (Ed.)
The Politics and Economics of Removing Subsidies on Petroleum Products in Nigeria
London & Abuja, Adonis & Abbey Publishers

Brasilia.Koplow, D. and Kretzman, S. (2010), 'G20 Fossil Fuel Subsidy Phase Out: A Review of Current Gaps and Needed Changes to Achieve Success,' Available at: http://earthtrack.net/files/uploaded_files/OCI.ET_.G20FF.FINAL_.pdf. (Accessed 24 January 2012).

Business News (2012), "GEJ Presents Nigeria 2012 Budget to National Assembly", Available at: http://businessnews.com.ng/2011/12/13/president-presents-nigeria2012-budget-to-joint/

Commission on Growth and Development (2008), *Growth report: strategies for sustained growth and inclusive development,* (Washington D. C., World Bank).

European Energy Agency (EEA) (2004), *Energy subsidies in the European Union: a brief overview,* (Copenhagen, EEA).

Filho, A. S. (2010), 'Growth, Poverty and Inequality: From Washington Consensus to Inclusive Growth,' DES Working Paper, No 100 ST/ESA/2010/DWP/100. Available at: http://www.un.org/esa/desa/papers/2010/wp100_2010.pdf

Ianchovichina, E.and Lundstrom, S. (2009), what is Inclusive Growth? Available at: http://siteresources.worldbank.org/INTDEBTDEPT/Resources/468980-1218567884549/WhatIsInclusiveGrowth20081230.pdf (Accessed: 14 January 2012)

International Energy Agency (IEA) Database (2012), Available at: http://www.iea.org/subsidy/index.html (Accessed 24 January 2012)

Kakwani, N. (2001), *Pro-poor growth and policies,* (Manila, Philippines, Asian Development Bank)

Kakwani, N., Khandker, S. and Son, H. H. (2004), 'Pro-Poor Growth: Concepts and Measurements with Country Case Studies', Working Paper, No. 1, International Poverty Centre,

Lang K. (2011), 'The First Year of the G-20 Commitment on Fossil-Fuel Subsidies: A Commentary on Lessons Learned and the Path Forward,' Available at: http://www.iisd.org/gsi/sites/default/files/ffs_g20_firstyear.pdf (Accessed 30 January 2012).

Chapter Six | Abdelrasaq Nal, in
Jideofor Adibe (Ed.)
The Politics and Economics of Removing Subsidies on Petroleum Products in Nigeria
London & Abuja, Adonis & Abbey Publishers

155

Nairaland (2012), "Aluko to Jonathan: Don't hide under subsidy," Available at: http://www.nairaland.com/827146/aluko-jonathan-don-t-hide (Accessed: 26 January 2012)

Organisation for Economic Cooperation and Development (OECD). (2006), *Promoting pro-poor growth: key policy messages*, (Paris, OECD Publishing)

Punch (2012), "Fuel subsidy and downstream oil sector deregulation," 03 January. Available at: http://www.punchng.com /editorial/fuel-subsidy-and-downstream-oil-sector-deregulation/

Ravallion, M. (2004), 'Pro-poor Growth: A Primer,' Policy Research Working Paper, No. WPS3242, Washington, D. C.: World Bank. Available at: http://papers.ssrn.com/sol3/papers.cfm?abstract_id=6 10283

Ravallion, M., and Chen, S. (2003), 'Measuring pro-poor growth', *Economic Letters*, vol. 78: 93-99.

Vanguard (2013), "FG finally removes fuel subsidy", 01 January. Available at: http://www.vanguardngr.com/2012/01/breaking-news-fg-removes-fuel-subsidy/

World Bank (2005), *Economic growth in the 1990s: learning from a decade of reform*, (Washington D. C., the World Bank).

World Bank (2009), *What is Inclusive Growth?* Available at:http://siteresources.worldbank.org/INTDEBTDEPT/Resources/4 68980-1218567884549/WhatIsInclusiveGrowth20081230.pdf.

Chapter Six | Abdelrasaq Nal, in
Jideofor Adibe (Ed.)
The Politics and Economics of Removing Subsidies on Petroleum Products in Nigeria
London & Abuja, Adonis & Abbey Publishers

156

CHAPTER SEVEN

The Macroeconomic Consequences of the Black Sunday in Nigeria

Efobi Uchenna, Osabuohien Evans & Beecroft Ibukun

Abstract

The Nigerian government's substantial reduction of fuel subsidy on Sunday, January 1, 2012 brought about shocks in some macroeconomic indicators. This paper examines the effect of the change in pump price of fuel on exchange rate, inflation and money supply. We developed an analytical model and the main variable that reflects the policy change is the pump price of fuel. Data was sourced from the Central Bank of Nigeria's website-monthly data on macroeconomic indicators from January 2009 to December 2012. Chow test for structural break and the Vector Autoregressive impulse response was used as tools for the estimations. We find a break point in the trend of the selected macroeconomic variables. Our findings have some policy implications which include the need for the government to consider timing when taking policy actions.

Keywords: Exchange rate, Fuel subsidy; Inflation; Money demand.

Introduction

In January 2012, the federal government of Nigeria substantially reduced the subsidy on fuel, which was supposed to be the government's effort in making household goods cheaper. Some have argued that the subsidy is one of the benefits many Nigerians derive from the country's 'cabalized' political regimes. This is because despite the transition to a democratic system of government, some of

---

Chapter Seven | Efobi Uchenna, Osabuohien Evans & Beecroft Ibukun, in Jideofor Adibe (Ed.)
The Politics and Economics of Removing Subsidies on Petroleum Products in Nigeria
London & Abuja, Adonis & Abbey Publishers

the dividends are yet to be realized. For instance, the total unemployment rate in 1998[18] was 7.6 percent, while in 2011; the rate was 23.9 percent (Akintoye, 2008; Central Bank of Nigeria, 2013).

The reduction of the subsidy has markedly increased the prices of fuel from $0.40/litre to about $0.61/litre presently. Explanations were put through to justify this action such as the mammoth cost on government to sustain the subsidy. For instance, in 2011, the fuel subsidy expense amounted to about 30 percent of the Nigerian government's expenditure, and 118 percent of the country's GDP. The value of the subsidy ($8 billion) was about four times the value of government expenditure on education ($2.2 billion) in 2011. This the Government noted as alarming and a distortion of priority as the funds could be used for other development projects. Some other explanations given for the reduction of the subsidy were that it will attract competition and be beneficial for private sector investment. These arguments were put forward because private individuals who will be interested in investing in the sector will be making a loss when their produce is sold at the subsidized price. Another argument is the issue of corrupt practices that have beclouded the administration of the subsidy. The probe that was carried out by the Federal government on the sector after the pronouncement revealed startling amounts of corruption.

Without much ado, this study is not intending to justify or castigate the government action, but to examine the behaviour of relevant macro-economic indicators as a result of the policy shock that the sudden shift in paradigm may elicit. We focused on three macroeconomic indicators- exchange rate, inflation rate and money supply. The main reason for these variables is that they can change within a short period of time. Beginning with exchange rate, the logic underlining the inclusion of this variable is that the subsidy reduction will affect the depreciation/appreciation of the domestic currency at

---

[18] 1998 marks the end of continuous military regime in Nigeria as democracy was fully restored 29th May, 1999.

Chapter Seven | Efobi Uchenna, Osabuohien Evans & Beecroft Ibukun, in Jideofor Adibe (Ed.)
The Politics and Economics of Removing Subsidies on Petroleum Products in Nigeria
London & Abuja, Adonis & Abbey Publishers

the world market. Changes in exchange rate depreciation will also affect the prices of imported goods: thus, a chain reaction.

We hypothesize that an increased rate of inflation will be caused by the subsidy reduction. This is because the general prices of goods will increase as a result of increased cost of production and transportation. Most manufacturing companies in Nigeria depend on privately generated power supply, which relies on fuel due to unreliable power supply from the national grid. Therefore a decrease in fuel subsidy is expected to increase the overhead cost on generating power supply. This increase will be transferred to the prices of goods and hence the resultant effect - inflation. The financial sector will also be affected as savings is expected to reduce because the demand for money will be higher to meet the increased prices of goods and services. All these occurrences have an effect on the level of poverty especially on households. Little wonder there was immediate reason for the populace to embark on a nation-wide strike and demonstration, cutting across almost all sectors of the national economy. In some cities like Lagos and Port Harcourt there was a disruption in productive activities for days.

This study therefore intends to find out if a structural break existed in the trend of these macroeconomic indicators, and how these variables responded to the shock caused by the fuel subsidy cutback. Our focus on these three variables is as a result of their relevance in defining the *well-being* of the economy (Australian Bureau of Statistics, 2012; Organization of Economic Corporation and Development-OECD, 2013). The remainder of the paper is structured as follows: the second section discusses the macroeconomic condition of Nigeria and the third draws the linkage between fuel subsidy diminution and poverty, linking the three macroeconomic variables. The fourth section reports the empirical results while the fifth section concludes with policy recommendations.

Chapter Seven | Efobi Uchenna, Osabuohien Evans & Beecroft Ibukun, in
Jideofor Adibe (Ed.)
The Politics and Economics of Removing Subsidies on Petroleum Products in Nigeria
London & Abuja, Adonis & Abbey Publishers

Stylized Facts: Macroeconomic Condition of Nigeria

To begin our discussion, it is worth observing the trend of the three variables of interest: exchange rate, inflation rate and money supply. We considered these variables using a trend plot for each of the 12 month periods between January 2009 and December 2012 in Nigeria. Figure 2.1 presents the trend plot for exchange rate in Nigeria and the Figure reveals that exchange rate has continued to rise consistently from the period of economic crisis denoted by the shaded plot. Focusing on the period of the fuel price reduction in 2012, we did not observe a significant improvement in the exchange rate but this period experienced the highest rate especially at the beginning period of the year.

Figure 2.1   Exchange Rate Regimes Per US Dollar (2009:01-2012:12)

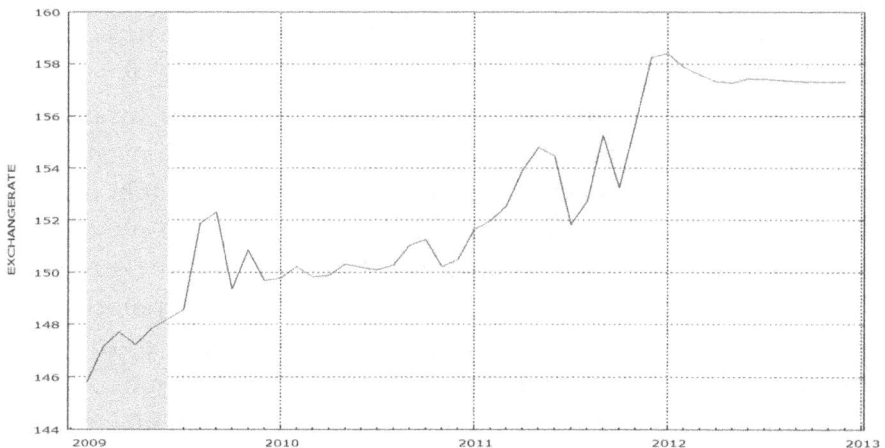

Source: Authors' compilation from CBN (2013)

We further examined the trend of inflation rate as presented in Figure 2.2. From the figure, we observed that inflation rate in Nigeria has consistently remained on a double digit. Considering the year of

Chapter Seven | Efobi Uchenna, Osabuohien Evans & Beecroft Ibukun, in
Jideofor Adibe (Ed.)
The Politics and Economics of Removing Subsidies on Petroleum Products in Nigeria
London & Abuja, Adonis & Abbey Publishers

160

the decrease in fuel subsidy, the trend reveals a different inflation rate trajectory. For instance, prior to 2012, inflation rate witnessed a consistent but undulating decreasing trend from the second month of 2010. However, in the later months of 2010, inflation rate began to rise but maintained a new path after 2012 as the value consistently remained above 11 percent compared to less than 10 percent in the preceding year.

Figure 2.2 Inflation Rate Regimes (2009:01-2012:12)

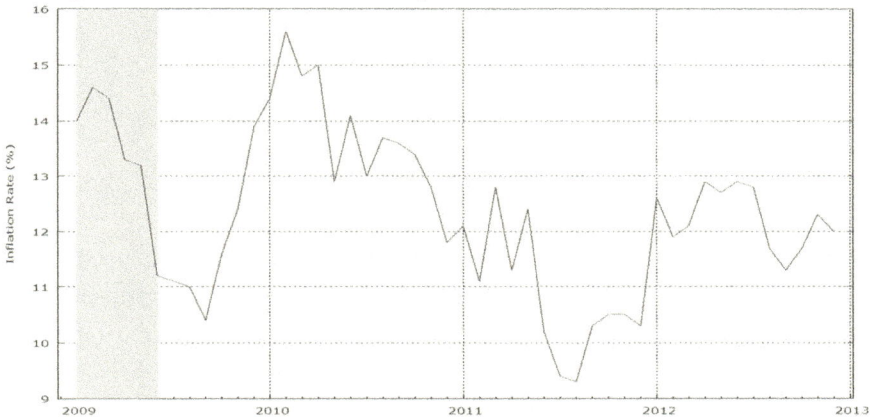

Source: Authors' compilation from *CBN (2013)*

Finally, the trend of money supply displayed in Figure 2.3 reveals that money supply has consistently remained on the increase between 2009 and 2012. However, we observed an initial shock immediately after the declaration of the reduction of subsidy in 2012, especially in the first month. This downward trend was maintained until later parts of the year when the trend regained its momentum and began to rise again.

Chapter Seven | Efobi Uchenna, Osabuohien Evans & Beecroft Ibukun, in
Jideofor Adibe (Ed.)
The Politics and Economics of Removing Subsidies on Petroleum Products in Nigeria
London & Abuja, Adonis & Abbey Publishers

Figure 2.3 Money Supply Regimes (2009:01-2012:12)

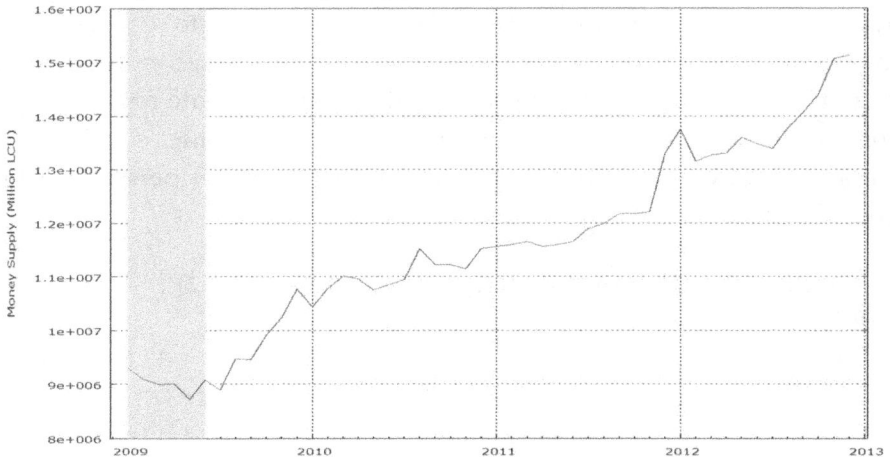

Source: Authors' compilation from CBN (2013)

From the trends, we have observed that these macroeconomic variables behaved differently especially in the early months of 2012. However, it is not so clear whether significant changes experienced by these variables were as a result of the fuel price increase and we will further test this in our empirical analysis.

Constitutional Error of the Black Sunday

Why the fuss about fuel subsidy reduction and the macro economy of Nigeria? The response to this question can be dilated but for focus, we will draw insight by linking the subsidy decrease to poverty. To begin with, we defined poverty using monetary measure: this implies that poverty can be viewed by how much a person earns, and by how much he/she spends on consumption (United Nations, 2010). This denotes that a person is as poor as his income and the level of his consumption. Relating this to the focus of this study, shrinking of fuel subsidy can *transcendentally* affect household poverty through the exchange rate, inflation rate and money supply. The linkage was illustrated in Figure 3.1.

Chapter Seven	Efobi Uchenna, Osabuohien Evans & Beecroft Ibukun, in
	Jideofor Adibe (Ed.)
	The Politics and Economics of Removing Subsidies on Petroleum Products in Nigeria
	London & Abuja, Adonis & Abbey Publishers

162

As can be seen in Figure 3.1, the Nigerian political leaders make policy decisions by first consulting professionals and special advisers and discussing with their executive council, which comprise of the ministers appointed to the cabinet. When decisions are reached through this process, the President sends a bill for approval to the legislature, which includes the Senate and House of Representatives. At this point, the senate deliberates over the proposed policy and when a decision is reached by a majority vote, the bill is passed and the President has to assent before implementation. These are denoted in Figure 3.1 by the straight lines running to and fro the political leader, public consultation, the executive council and the legislature.

Various documented and undocumented reports reveal that the substantial withdrawal of the petroleum subsidy by the President, tagged 'Black Sunday' did not follow this process. Though the initial consultations and meetings with the executive council were ongoing, the decision to remove the subsidy was not approved by the legislature or ever mentioned in the 2012 national budget. This represents a somewhat 'betrayal', as well as an undemocratic move by the Government, showing little regard for both the National Assembly and the Nigerian people as a whole.

This unconstitutional action will have an immediate shock on the economy. This is especially so with regards to prices of fuel dependent economic activities such as transportation and production. Considering transportation, a 100 percent price increase was observed as prices of transportation doubled. Similarly, the pump price of fuel increased at the instance of the announcement made by the President of the subsidy reduction. Due to this increase, production processes became expensive as most factories generated their own electricity by using generator sets.

The output from the production process can be channeled to either the external context in the form of export or can be internally consumed. This is denoted by the broken line flowing from production to the two sectors. The ability of domestic firms to compete with their foreign counterparts both at home and abroad is

Chapter Seven | Efobi Uchenna, Osabuohien Evans & Beecroft Ibukun, in
Jideofor Adibe (Ed.)
The Politics and Economics of Removing Subsidies on Petroleum Products in Nigeria
London & Abuja, Adonis & Abbey Publishers

influenced by the relative prices of domestic and foreign produced goods (Schembri, 1989). We infer that by the fuel price subsidy reduction, there will be an increase in the production costs of goods and services, leading to an increase in the prices of goods and services bringing about a negative effect for both the internal and the external sectors. Based on this, the demand for goods will dwindle.

Chapter Seven | Efobi Uchenna, Osabuohien Evans & Beecroft Ibukun, in
Jideofor Adibe (Ed.)
The Politics and Economics of Removing Subsidies on Petroleum Products in Nigeria
London & Abuja, Adonis & Abbey Publishers

Figure 3.1:   The Constitutional Error of the Black Sunday

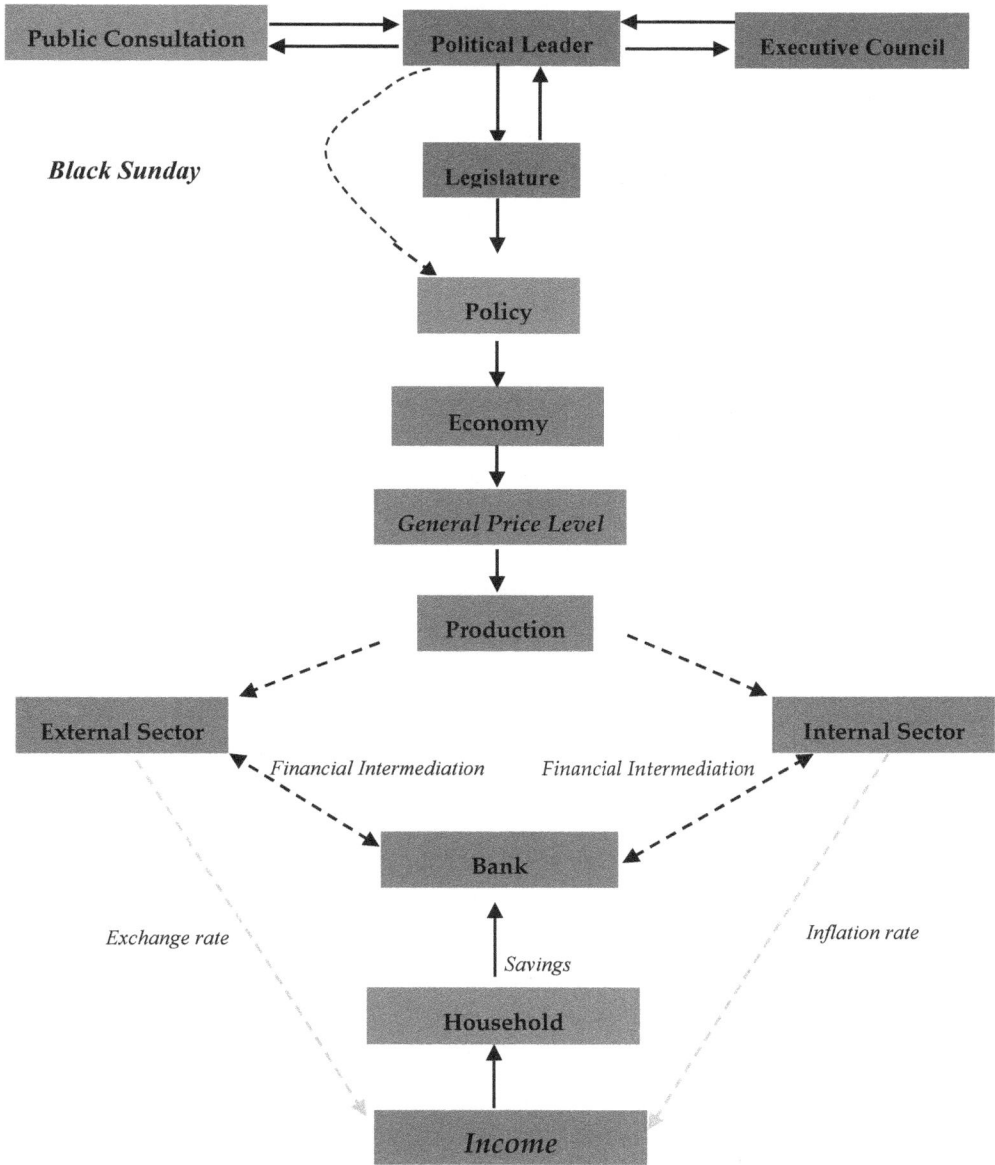

Source: Authors'

Chapter Seven | Efobi Uchenna, Osabuohien Evans & Beecroft Ibukun, in
Jideofor Adibe (Ed.)
The Politics and Economics of Removing Subsidies on Petroleum Products in Nigeria
London & Abuja, Adonis & Abbey Publishers

At the level of the external sector, an increase in production costs will reduce firms' profitability and the ability of domestic industries to compete in the world market. This in turn will have adverse effects on the exchange rate of the country as well as on the trade balance/balance of payment. Local currency depreciation will occur as the value of the currency begins to fall because customers abroad would have lower demand for the local currency due to higher prices of goods. They will rather seek for competitive goods that are cheaper.

On the domestic platform, output channeled to local consumption will equally bear the high price caused by an increase in production costs. This will raise the general price equilibrium in the economy - inflation, which erodes the purchasing power of the households, taking into account the fact that the value of household incomes will depreciate (Olayiwola, Osabuohien and Efobi, 2012). Apart from the increase in the prices of goods, prices of services that are fuel dependent will also increase. This shock will affect the consumption pattern of the households as the increased prices of goods will bring a shock to the household income as their incomes do not increase at the same rate as prices (Bacon, 2005; Energy Management and Assistance Program, 2005). The fuel subsidy reduction, represented a 49 per cent increase in the pump price of fuel, however, the general incomes of households did not increase rapidly by this percentage. For instance, the then minimum wage of about 113 USD (₦18,000) for civil servants has not increased. Based on this, we expect households to be worse off as their savings will reduce, causing a sharp decrease in the general savings of the country due to higher demand for money and dis-incentive towards savings. We expect this to affect the money supply in the country. This is because the money supply function performed by the Central Bank will be affected by the extent of the general savings. The resultant effect will be a reduction in the liquidity of the banking sector, thereby limiting the amount of credit available to businesses that may be engaged in production.

Chapter Seven | Efobi Uchenna, Osabuohien Evans & Beecroft Ibukun, in Jideofor Adibe (Ed.)
The Politics and Economics of Removing Subsidies on Petroleum Products in Nigeria
London & Abuja, Adonis & Abbey Publishers

166

To empirically validate our proposition, we will examine the presence of a structural break in January 2012, in relation to the trend of these identified macroeconomic indicators. This will be further corroborated by the speed of response of these indicators to the shock from fuel prices. This study is interested in three macroeconomic indicators: exchange rate, inflation rate and money supply, because these indicators have significant macroeconomic consequences especially on trade, balance of payment, poverty and the banking sector development (Ojapinwa and Ejumedia, 2012; Olayiwola, Osabuohien and Efobi, 2012). The main dependent variable that reflects the policy change- subsidy reduction is the fuel price per barrel. Following anecdotal evidence, when the fuel subsidy was reduced, the barrel price was affected. Data for this study was sourced from the Central Bank of Nigeria website for the period January 2009 to December 2012. Chow test for structural break and the Vector Autoregressive Impulse Response were used as tools for the estimations.

Empirical Discussion

To determine the impulse response of the three macroeconomic variables on fuel subsidy cutback, we used the Vector Autoregressive-VAR function. The VAR function is represented in equation 4.1:

$$Fp_t = \alpha + \sum_{j=1}^{k} \beta_j Fp_{t-j} + \sum_{j=1}^{k} \gamma_j \, Macro - economic \, Variables_{t-j} + e \quad (4.1)$$

From this model, we are interested in the reaction of the macroeconomic variables as a result of shocks or innovations from the stochastic error term e, due to changes in Fp (fuel price as a result of fuel subsidy reduction). We imply that if e increases by a value of one standard deviation, Fp will be affected in its contemporaneous and future values, such that the effect of the change traces out over a period of time. Likewise, the macroeconomic variables react as well to a shock in e as a result of changes in Fp and the effect fades with time.

Chapter Seven | Efobi Uchenna, Osabuohien Evans & Beecroft Ibukun, in
Jideofor Adibe (Ed.)
The Politics and Economics of Removing Subsidies on Petroleum Products in Nigeria
London & Abuja, Adonis & Abbey Publishers

Put differently, the effect of a shock in the current year will 'disappear' with time.

We performed the impulse response and reported the output in Figure 4.1. From the Figure, fuel price responded drastically to a shock in its own price as a result of the diminution of the subsidy. The response was negative implying a consistent rise in fuel price and it maintained this negative regime without adjusting all through the period. As expected, the other three macroeconomic variables responded negatively at the instance of a shock on fuel price as a result of the subsidy cutback. Money supply reacted negatively but adjusted after the first period. While still unfavorable, it maintained a constant negative trend all through the period. Inflation rate did not readjust after the first negative shock but sustained its continuous undesirable trend all through the period. Exchange rate followed a similar trend but began to adjust to equilibrium after the third period and maintained a negative trend from the sixth period onward.

Figure 4.1 Impulse Responses of Variables to a Shock in Fuel Price

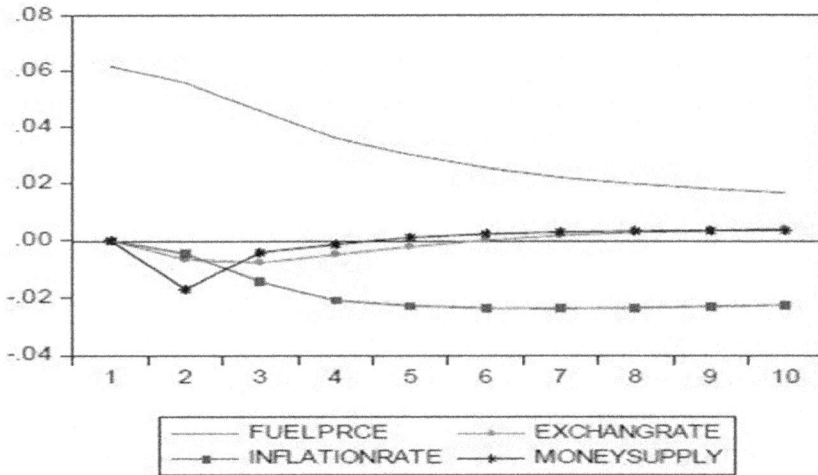

Source: Authors'

Chapter Seven	Efobi Uchenna, Osabuohien Evans & Beecroft Ibukun, in Jideofor Adibe (Ed.) *The Politics and Economics of Removing Subsidies on Petroleum Products in Nigeria* London & Abuja, Adonis & Abbey Publishers

The impulse response shown in Figure 4.1 depicts the reaction of the macroeconomic variables to a shock in fuel price. From the Figure, a policy shock in fuel price will bring about an initial drop in exchange rate and money supply but after a while, the effect begins to fade away. Inflation rate consistently dropped at the first period of the shock and maintained the trend all through the period. However, we are more interested in finding out if a significant structural break existed in the trend of the macroeconomic variables using January 2012 as the structural breakpoint date.

The idea of estimating the chow breakpoint is to fit the relationships between the macroeconomic variables and fuel price and to observe the existence of structural differences in the relationships. Thus, we will find out whether the macroeconomic variables reacted to the fuel price shock significantly after the structural breakpoint date. Our result was presented in Table 4.1 below and we observed that a significant structural change occurred after January 2012 in all three relationships.

Table 4.1: Chow Breakpoint-January 2012

Fuel price ⟶Exchange rate	7.1775	13.5241
	(0.0021)	(0.0012)
	52.8473	58.7716
Fuel price ⟶Inflation rate	(0.0000)	(0.0000)
	5.8376	11.2966
Fuel price ⟶ Money supply	(0.0058)	(0.0035)

Note: Values in parenthesis are probability values

From the Table, the exchange rate, inflation rate and money supply in Nigeria all followed a different regime after the pronouncement of the reduction of the fuel subsidy by the government in January 2012. This supports our proposition that the

Chapter Seven | Efobi Uchenna, Osabuohien Evans & Beecroft Ibukun, in
Jideofor Adibe (Ed.)
The Politics and Economics of Removing Subsidies on Petroleum Products in Nigeria
London & Abuja, Adonis & Abbey Publishers

169

recent government policy on fuel subsidy has serious macroeconomic implications.

Summary and Conclusion

This study investigated the existence of a structural break in the relationship between exchange rate, inflation rate, money supply and fuel price holding January 2012 as the structural breakpoint. The Chow breakpoint test was used and the impulse response test was also performed to establish the behaviour of the variables to a shock in the domestic fuel price. We established that there was a sharp reaction of the macroeconomic variables of interest to fuel subsidy reduction considering the structural break and impulse response function. The implication of this finding is that fuel price is a very sensitive variable which can alter the macroeconomic equilibrium of an economy. Therefore the Government should have an understanding of this when deliberating on future policies. This will help in reducing socio-economic tensions of policies as the populace will need some time to adjust to policy changes. Despite the Government's 'good' intentions for policy change such as the substantial cut in the fuel subsidy, when the process lacks proper timing, the macroeconomic consequences can be gruesome.

References

Akintoye, R. (2008), 'reducing unemployment through the informal sector: A case study of Nigeria'. In *European Journal of Economics, Finance and Administrative Sciences*, 11.

Australian Bureau of Statistics (2013), 'Key Economic Indicators,' Available at: *http://www.abs.gov.au/ausstats/abs@.nsf/mf/1345.0. Accessed 5th May, 2013*

Bacon, R. (2005), *the impact of higher oil prices on low income countries and on the poor*, (Washington DC, World Bank Energy Sector Management Assistance Programme).

Chapter Seven | Efobi Uchenna, Osabuohien Evans & Beecroft Ibukun, in
Jideofor Adibe (Ed.)
The Politics and Economics of Removing Subsidies on Petroleum Products in Nigeria
London & Abuja, Adonis & Abbey Publishers

170

Central Bank of Nigeria (2012), *Statistical Bulletin*. (Abuja: Central Bank of Nigeria).

Central Bank of Nigeria (2013), *website*, Available at: *http://www.cenbank.org.ng*. Accessed 5th May, 2013"

Energy Management and Assistance Program (2005), 'The Vulnerability of African Countries to Oil Price Shocks: Major Factors and Policy Options', *Esmap Report 308/05.*

Organisation for Economic Cooperation and Development-OECD (2013), 'Main Macroeconomic Indicators'. Available at: *http://stats. oecd.org/mei/default.asp?rev=4. Accessed 5th May, 2013*

Ojapinwa, T. V. and Ejumedia, P. E., (2012), 'The industrial impact of oil price shocks in Nigeria', in *European Scientific Journal*, 8(12).

Olayiwola, W., Osabuohien, E., and Efobi, U., (2012), 'Inflation and Financial Development in Nigeria', in Olaniyan, O., Lawson, A. O., and Alayande, B., (eds.), *Financial Sector Issues and Economic Development in Nigeria*, (Ibadan: University Press).

Schembri, L., (1989). Export Prices and Exchange Rates: An Industry Approach, Available at: *http://www.nber.org/chapters/c6180. Accessed 5th May, 2013"*

| Chapter Seven | Efobi Uchenna, Osabuohien Evans & Beecroft Ibukun, in Jideofor Adibe (Ed.) *The Politics and Economics of Removing Subsidies on Petroleum Products in Nigeria* London & Abuja, Adonis & Abbey Publishers |

CHAPTER EIGHT

Petroleum Products Pricing and the Manipulation of Oil Subsidy in Nigeria

Adadu Yahaya Abdullahi, Usman Abu Tom & Hafsat Kigbu

Introduction

Nigeria is blessed with vast quantities of oil, and is currently the sixth largest oil exporter among the Organization of Petroleum Exporting Countries (OPEC). Since the discovery of oil in Nigeria billions of dollars have been generated as revenue. Apart from being a major source of revenue to the Nigerian economy, oil constitutes not only a reservoir of energy for the nation but also serves as a major determinant of its foreign policy. Despite the overwhelming influence of the petroleum industry in terms of its contribution to export earnings and government revenues in Nigeria, strong and persistent sentiments are always expressed by vast number of Nigerians about the dismal state of affairs, particularly the severe disruption in fuel supplies and intermittent upward price reviews that have characterized the Nigerian petroleum industry. Indeed, only politics elicits as much passion and tension as the issue of petroleum products pricing in Nigeria. This is because energy is fundamental to human life and plays a vital role in economic prosperity of the country and the overall well-being of the people.

The supply of petroleum products remains an acid test for successive regimes as the passion and tension that usually characterize petroleum discourse are reactions to the inexplicable deprivations and suffering of Nigerians amid plenty and abundant of petroleum products. In 1998, the regime of General Abdulsalami Abubakar moved towards deregulation by allowing the oil marketing

Chapter Eight | Adadu Y. Abdullahi, Usman A. Tom & Hafsat Kigbu, in
Jideofor Adibe (Ed.)
The Politics and Economics of Removing Subsidies on Petroleum Products in Nigeria
London & Abuja, Adonis & Abbey Publishers

173

companies to import fuel directly. In the past this was the exclusive preserve of the NNPC. However, importation was unattractive to the major marketers due to the local fixed price regime.

In 2003, however, the regime of Chief Olusegun Obasanjo announced a full deregulation of petroleum downstream sector. According to Obasanjo (2003), deregulation of the petroleum downstream sub-sector was a way of opening up the petroleum industry to competition and enhanced efficiency. But those against this argue that, deregulation leads to steady increase in the price of petroleum products which will concomitantly aggravate the poverty level of poverty in the country, as it will mean that more people will no longer be able to meet their basic needs of food, clothing, shelter, education etc. The primary objective of this paper is to examine the issues involved in pricing of petroleum products and removal of subsidies, with particular reference to the period 1999-2012, with a view to capturing the phenomena of fuel crises in Nigeria. . After the introduction, Section Two gives an overview of the petroleum industry in Nigeria while Section Three is devoted to the historical analysis of petroleum products pricing from 1999 to 2012. The administration and manipulation of oil subsidies by the Nigerian state is examined in section four. Section five concludes the paper.

An Overview of the Nigerian Petroleum Industry

Historically, the actual period of the first discovery of petroleum in Nigeria has remained controversial. This is because some prominent scholars based the formation of oil industry on contemporary writings and oral postulations, while others traced it to the period of colonialism.The very first attempt to discover oil in Nigeria was made in 1908 by a German company called the Nigerian Bitumen Corporation, at Araromi, presently in Ondo State. Shell Development Company of Nigeria, a subsidiary of Shell d'Archy and British Petroleum Company (later Royal Dutch Shell BP) joined exploration

Chapter Eight | Adadu Y. Abdullahi, Usman A. Tom & Hafsat Kigbu, in Jideofor Adibe (Ed.)
The Politics and Economics of Removing Subsidies on Petroleum Products in Nigeria
London & Abuja, Adonis & Abbey Publishers

174

in 1937 and first discovered petroleum in commercial quantity in 1956 at Oloibiri in the present Bayelsa State. By 1961 other companies such as Total Agip and Elf joined Shell and together produced about 5,100 barrels of crude oil per day (Roberts, 1998:11). Since then, the production of petroleum started increasing in Nigeria.

Nigeria is ranked as the number one oil-rich country in Sub-Saharan Africa and, the sixth largest within the Organization of Petroleum Exporting Countries (OPEC) and has a technical capability to produce about 12.6 million barrels per day. Oil export accounts for 98.5% of Nigeria's total exports. This is a clear indication that Nigeria's economy absolutely depends on production and sales of crude oil. The discovery of oil in the mid-1950s has good potentials for the delivery of good governance, social justice, economic independence and political stability for the overall development of the country.

But the promise has hardly been realized in part, because the oil industry has been closely associated with instability. It has caused spiral inflation and made Nigerians to import and consume all kinds of goods, including those that could be produced in the country. It promoted structural imbalance, institutionalised the culture of monolithic economy, propelled high rate of inflation, and has displaced agriculture as a major contributor of government revenue, expenditure and national growth.

Presently, Nigeria has four refineries of international standard with a total installed refining capacity of 445,000 barrels per day and refines part of its crude oil domestically for internal consumption. The oldest is the Port Harcourt refinery which was commissioned in 1965. The Warri, Kaduna, and Port Harcourt refineries were commissioned in 1978, 1980 and 1989 respectively. All these refineries produce the normal range of petroleum products such as Liquefied Petroleum Gas (LPG), Premium Motor Spirit (PMS), Kerosene and Automotive Gas Oil (AGO).

Chapter Eight | Adadu Y. Abdullahi, Usman A. Tom & Hafsat Kigbu, in
Jideofor Adibe (Ed.)
The Politics and Economics of Removing Subsidies on Petroleum Products in Nigeria
London & Abuja, Adonis & Abbey Publishers

Over the years, Nigeria derived revenue from oil through numerous ways, exploration of crude oil, petroleum profit, tax, royalties, rent and price adjustment. Oil revenue has been the determinant of government expenditure in Nigeria. According to Ajidani and Sani (2005), since the big oil price increase in 1980, oil has provided over 80% of Nigeria's export earnings and on the average over 90% of government revenue between 1980 and 2000. It has been the major determinant of employment, improved wages, social amenities and infrastructure.

It is instructive to remark that despite Nigeria been the largest producer of crude oil in Africa and sixth in the world, she is faced with petroleum crises. These crises have been caused by several factors. The plausible determinants are attributed to the vastly expanded energy consumption, irregular maintenance of the refineries which often result to total breakdown, large scale smuggling of petroleum products to neighbouring countries enhanced by corrupt government officials and the role played by saboteurs behind the constant vandalization of pipelines due to the divide and rule strategy used by both government and oil firms to avoid payment of legitimate claims by oil communities and the incessant strikes by oil workers, mainly those under the aegis of National Union of Petroleum and Natural Gas Workers (NUPENG) and the Petroleum and Natural Gas Senior Staff Association of Nigeria (PENGASSAN).

The above outlined factors which have fuelled of petroleum products crises in Nigeria; clearly contribute to the shortage of supplies of petroleum products which invariably results in massive importation to lessen the suffering of the population. Chronic fuel shortages erupt in the country. With the fuel hike also came the official announcement of the era of the deregulation of the oil marketing industry.

Nigerians tend to be in favour of regulation as a way of curbing the excess of petroleum marketers. According to (Oluleye, 2005) the

Chapter Eight | Adadu Y. Abdullahi, Usman A. Tom & Hafsat Kigbu, in Jideofor Adibe (Ed.)
The Politics and Economics of Removing Subsidies on Petroleum Products in Nigeria
London & Abuja, Adonis & Abbey Publishers

main factors that affect the demand for regulation are consumer surplus per buyer, number of buyers, producers surplus and number of firms. The larger the consumer surplus per buyer that results from regulation, the greater is the demand for regulation by buyers. Also, as the number of buyers increase, so does the demand for regulation. It is as a result of the above instance that the federal government committee on the review of the petroleum products supply and distribution recommended the establishment of the Petroleum Products Pricing and Regulatory Agency. Its (establishment) Act 2003 was enacted by the National Assembly. This gave birth to PPPRA, a tool used by government to seek for appropriate pricing of petroleum products. This agency heralded an era of incessant increases in prices of petroleum products in Nigeria. Egwaikhide et al (2004) argued that the traditional explanation for all these increases revolves around the following: to remove the ever-present government 'subsidy' on petroleum products and make them more readily available.

Petroleum Products Pricing In Nigeria 1999-2012

Petroleum products constitute a major input in the production of various goods and services, consequently, changes in supply affects the cost of production and, therefore, price level (Egwaikhide, 2004). Before1973, petroleum products pricing was not uniform in Nigeria. The retail prices of petroleum products were dependent on the point of sale, relative to the only primary distribution depot then, at the Shell-BP refinery, Alesa Eleme, near Okrika, near Port Harcourt, River State. In October, 1973, the military government of General Yakubu Gowon decreed uniform pricing of refined petroleum products for the Nigerian market. Subsequently, the Petroleum Equalization Fund (PEF) Decree No 9 of 1975 was promulgated (Yamma, 2011). It should be stated that before the deregulation policy of the downstream oil sector in September 2003, PEF provided for the uniform pricing of all

Chapter Eight | Adadu Y. Abdullahi, Usman A. Tom & Hafsat Kigbu, in
Jideofor Adibe (Ed.)
The Politics and Economics of Removing Subsidies on Petroleum Products in Nigeria
London & Abuja, Adonis & Abbey Publishers

177

petroleum products throughout Nigeria, irrespective of distance from the point of production.

The fuel crises phenomenon in Nigeria is manifested in the form of incessant hike in the prices of petroleum products, mainly the PMS. A year after inauguration of former president Obasanjo's civilian administration precisely on June 1st, 2000, Jackson Gaius Obaseki former Group Managing Director of the Nigerian National Petroleum Corporation (NNPC) announced the new pump price of the products which were N30 per litre (PMS), N29 per litre (Diesel) and N27 per litre (Kerosine) respectively. However, following a mass protest by labour unions, students, civil societies and the general public, government reached a compromise with representatives of labour, civil societies, the price of these products were forced down in June 3, 2000 to N25, N21 and N17 per litre for PMS, diesel and kerosene respectively. In June 13, there was another reduction in the price of PMS due to continued resistance by labour, which brought down the price to N22 per litre.

The increment announced at the peak of new Year festivities in January 1st, 2002 saw the price of PMS increased from N22 to N26 per litre, diesel from N21 to N26 per litre and Kerosene (the 'poor persons fuel') from N17 to N24 per litre . As noted by Egwaikhide et al (2004), with the fuel price hike also came the official announcement of the era of liberalization of the oil marketing industry in Nigeria. This liberalization policy in the oil sector necessitated the emergence of the PPPRA with Chief Rasheed Gbadamosi as its first Chairman. This committee heralded an era of incessant increase in prices of petroleum products in Nigeria.

It is evident, that between 2000 and 2012, the prices of petroleum products have been increased nine times. During Obasanjo's eight years rule as a democratically elected president 1999 to 2007, the prices were changed eight times: June 1, 2000; January 1, 2002, June 20, 2003; October 1, 2003; May 29, 2004; January 2005; August 2005 and May 25, 2007. It is clear that the administration of Olusegun

Chapter Eight | Adadu Y. Abdullahi, Usman A. Tom & Hafsat Kigbu, in
Jideofor Adibe (Ed.)
The Politics and Economics of Removing Subsidies on Petroleum Products in Nigeria
London & Abuja, Adonis & Abbey Publishers

Obasanjo removed oil subsidy six times; the price of petrol (PMS) rose from N20 to N75 per litre (Late President Yar'Adua later reversed it to N65).

The Goodluck Jonathan administration on the first day of January, 2012 shocked Nigerians when he suddenly hiked prices of PMS to N147 per litre. This act led to the doubling of transport fares in the country. This was particularly painful for hundreds of thousands of Nigerians who had gone to their home towns and villages for the Christmas and New Year holidays; most of them were stranded due to the fact that their budgets no longer added up and they had to either borrow or dip into their savings to return to their various places of work and abode. In response to the hike in price, Nigerians spontaneously broke out into streets protesting against the policy. The Nigerian Labour Congress (NLC) and other organized pressure groups embarked on a nationwide strike that grounded the country and threatened the entire economy and the country's nascent democracy. The regime had to announce a reduction in the price of PMS to N97 per litre. The frequent argument for all these increases by various regimes revolved around removing the ever present government 'subsidy' on petroleum products and make them more readily available; and money saved from the removal of oil subsidy would be used to develop and rebuild the much needed infrastructure in the country. They further argued that the increase would prevent smuggling and illegal exportation of refined products to neighbouring countries where such products are much more expensive (Ukpong, 2004).

Chapter Eight | Adadu Y. Abdullahi, Usman A. Tom & Hafsat Kigbu, in
Jideofor Adibe (Ed.)
The Politics and Economics of Removing Subsidies on Petroleum Products in Nigeria
London & Abuja, Adonis & Abbey Publishers

Table 1: Overview of petroleum products pricing in Nigeria N/ litre 1999-2012

DATE	PRICE PER LITRE (N/K)	REGIME	INCREASE (%)
Jan 6, 1999	20	Gen. A. Abubakar	-20
Jun 1, 2000	30	Olusegun Obasanjo	50
Jun 8, 2000	25	" "	-16.66
Jun 13, 2000	22	" "	-12
Jan 1, 2002	26	" "	18.2
Jun 20, 2003	40	" "	53
July 9, 2003	34	" "	-15
Oct 1, 2003	42	" "	23.5
May 29,2004	49.9	" "	18.81
Dec 21, 2004	48	" "	-13.81
Jan 1,2005	50.5	" "	5.2
Aug ,2005	65	" "	28.71
May 25,2007	75	" "	
Jun 24,2007	65	Umaru Yar'adua	-25.69
Jan 1,2012	141	Goodluck Jonathan	110
	97	" "	-70

Source: This day newspaper, 2 January 2012.

The history of the increases in prices of petroleum products in Nigeria can be linked to the state's desire to implement the World Bank and IMF proposal of deregulating the economy and removing government's involvement in key areas of the economy and channelling same to other needing sectors like health, education and infrastructure. Hence, the government's claim of removing subsidy from petroleum products that translates to so much money is usually the government's way of achieving increases in the pump prices of

Chapter Eight | Adadu Y. Abdullahi, Usman A. Tom & Hafsat Kigbu, in Jideofor Adibe (Ed.)
The Politics and Economics of Removing Subsidies on Petroleum Products in Nigeria
London & Abuja, Adonis & Abbey Publishers

180

petroleum products. The argument by government has always been that so much money is being expended on the subsidies that end up benefiting very few people in the society. And that such money will be more useful to citizens in health, education and infrastructure. The Labour and Civil Society's reaction to this has been of two brands, first, it constituted by the group that feel subsidy these are rights of citizens and that they represent the benefits of the common Nigerian from the God-given resources that is endowed naturally in the country. The second group is comprised of a section of the civil society group that government has no subsidy to remove from the prices of petroleum products, and that Nigerians are already paying the full costs of the petroleum products, from the landing costs to the distribution costs. For this second group, government is only engaging in increases in petroleum products to continue to fund its ever increasing budgets of political campaigns and the huge cost of running the enlarged budget of the presidential system of government that the political elites insists on pushing along. The group has also argued that the price increases are manifestation of the government's complicity in the fight against corruption that pervades the down-stream sector of the petroleum industry.

The above confrontation between the government, the civil society and labour usually leads to strikes, lock-out and protests across the streets of the cities of the country as was witnessed in the January, 2012 nationwide anti-fuel subsidy protest. The events are usually followed by calls for caution by well-meaning Nigerians, and then it translates into dialogue between the government and labour and civil society, and then an eventual settlement on a certain price. This has always been the trend such as the January, 2012 experience that placed the PMS Price at 141 naira per litre and eventual settlement was made at 97 naira per litre.

Chapter Eight	Adadu Y. Abdullahi, Usman A. Tom & Hafsat Kigbu, in
	Jideofor Adibe (Ed.)
	The Politics and Economics of Removing Subsidies on Petroleum Products in Nigeria
	London & Abuja, Adonis & Abbey Publishers

The State and the Manipulation of Oil Subsidy in Nigeria

The issue of fuel subsidy has been on the front burner of every successive administration of the Nigerian state. From General Ibrahim Babangida in 1985 to the present government of Goodluck Jonathan. Usually, each government would claimed that there has been huge fuel subsidies that made other infrastructural facilities to be compromised. In line with this, Roberts (1998) has argued that, the heavy subsidy is a barrier to sustainable growth. In the same vein Okonjo-Iweala, Nigeria's finance minister, stated that the 2012 budget excluded fuel subsidy because it has become very expensive for the government to continue to subsidise due to the effect of the current global economic crises on the world economy. She further added that the government spent whooping N1.3 trillion as fuel subsidies.Subsequently,the federal government included the subsidy fund in the 2012 supplementary appropriation budget.

The argument further points to the very dangerous reality that in fact it is not at all about large sums deployed to subsidy but that a few Nigerians have constituted themselves into a clique that is benefiting heavily from this leakage called subsidy as evident in the House of Representatives investigation team under the auspices of Honorable Faruq Lawal's adhoc committee on subsidy probe in 2012. This argument further means that huge sum deployed to subsidy is merely funding a few wealthy citizens who are feeding fat on the nation while the rest of the population is in dire need of virtually all critical infrastructure. In the same vein Sanusi Lamido Sanusi , the CBN governor argued that;

> What we have in our hands is subsidy of cost of consumption. When you subsidise cost of Consumption, it means you are subsidizing cost of production in those foreign countries. You are producing employment for their people. For me, I think the agitation the labour and the civil society should spear is how to subsidize cost of production in Nigeria so that there would be more jobs for our people. As it is, Nigeria is not subsidizing fuel

Chapter Eight | Adadu Y. Abdullahi, Usman A. Tom & Hafsat Kigbu, in
Jideofor Adibe (Ed.)
The Politics and Economics of Removing Subsidies on Petroleum Products in Nigeria
London & Abuja, Adonis & Abbey Publishers

182

for the masses but paying rents to few people who are milking the nation: (Daily Trust, 23 December 2011).

Subsidy removal is normally one of the elements in a typically IMF/World Bank supported structural adjustment programme. In 2002, Sanjeev Gupta and a few colleagues in the IMF wrote a working paper on domestic petroleum pricing in oil producing countries. In that paper they argued that;

> Petroleum product prices were heavily regulated. Domestic petroleum prices were below international Price and this implied foregone revenue. The foregone revenue constitutes a hidden subsidy for its citizens (Price-gap methodology). This hidden subsidy benefited high income more than the low income group. The solution was to pass international price into domestic market and increase fuel price (Gupta, 2002:16)

They recommended the imposition of international prices on the domestic petroleum products and the removal of fuel subsidies. They further advised on identifying political opponents of the fuel subsidy removal program, how to do a publicity campaign , set up a programme aimed at using the money generated , how to time the subsidy removal, how to make promises of transport buses, education, health, roads and give money to the poor if necessary. Their paper is the blueprint of the fuel subsidy removal programme that the Nigerian government is unleashing on Nigerians today.

At this point we ask, what does fuel subsidy mean and was subsidy ever in place in Nigeria? And who benefited from it? Subsidy simply means benefit given by government to individuals or businesses whether in form of cash, tax reduction or by reducing the cost of goods and services. The purpose of subsidy is to help individuals and businesses to purchase or acquire essential goods and services that they may not be able to afford, under normal circumstances. Subsidy is a government programme created to reduce how much Nigerians have to pay for petroleum products, which include: Premium Motor Spirit, Automotive Gas Oil and Dual

| Chapter Eight | Adadu Y. Abdullahi, Usman A. Tom & Hafsat Kigbu, in Jideofor Adibe (Ed.)
The Politics and Economics of Removing Subsidies on Petroleum Products in Nigeria
London & Abuja, Adonis & Abbey Publishers |

Purpose Kerosene, and to protect the citizens from crude oil volatility on the international market.

The expenditure of the government subsidies on petroleum products in Nigeria remains debatable. According to the presidency, subsidies exist when we compare the local price here with what is obtainable at the international market. Since the selling price here is lower than the price at the international market, therefore subsidy exists. In other words, the government sees subsidy in terms of maintenance of the price parity with the rates in other countries using the dollar index.

Abubakar (2011) has argued that:

> The first mistake we made is to consider the export price of a barrel of crude as our base price. We should not. How much we could realize if we had exported the crude is false and misleading. The amount earmarked for domestic refining is, legally, not exportable except if we intend to use the proceeds to import our shortfall. The selling price of say US$75 dollars is thus merely an opportunity cost (Abubakar, 2011:64).

He further stated that;

> You don't grow yam and then tell your children let us sell the yam, and then buy prepared pounded yam, because the selling–current price of yam is very attractive. The pounded yam you kept buying would end up costing You more than the amount you realised from selling the raw yam (Ibid).

It is very clear that the government is not sincere to its citizens. OPEC gave a concession to the Federal Government to exceed its quota by 445,000 barrel a day to provide for the domestic needs of Nigerians. But the government, because it finds it lucrative, crippled the four refineries through neglect so as to continue selling this allocation internationally and making huge profits while using part of the proceeds to import refined products in a very non-transparent, where the importers bring in cheap and loaded petrol with no special additives. Even the Department of Petroleum Resources (DPR) has

Chapter Eight   Adadu Y. Abdullahi, Usman A. Tom & Hafsat Kigbu, in
Jideofor Adibe (Ed.)
The Politics and Economics of Removing Subsidies on Petroleum Products in Nigeria
London & Abuja, Adonis & Abbey Publishers

184

accused them of lifting from Nigeria, going to the high-seas and returning to claim they imported the product and are entitled to subsidy claims.

The unprecedented rate at which the cost of subsidy keeps growing shows that something is wrong somewhere in its administration by the NNPC and the PPPRA. The inconsistency in the amounts deducted for subsidy of petrol led both the states and federal governments to call for the stopping of such deductions. The state contends that there were discrepancies in the subsidy deductions made by NNPC and PPPRA. There is no doubt that the problems of fuel subsidy are not restricted to revenue loss but lack of transparency in the management of even the subsidy fund.

The corruption in subsidy management is one the reasons why the leaders of the country have not done much to make the refineries in the country work. If the refineries are in working condition, most of the cost incurred in importing fuel will stop.

In an interview in the New Magazine (October 9, 2000) the late Professor Sam Aluko stated:

> How can we have four refineries in the country and then the four would breakdown at the same time? Even when we tried to award the contract to Total, they kept telling us that Total could not do it. Total has seventeen refineries around the world; they are all working. It's only our refineries that they cannot put in order. You know something is wrong somewhere, Corruption (News Magazine,October 9,2000:26).

Aluko's argument then was to stop the importation of oil products, so that the problem and contestation surrounding subsidy could be tackled. Again the expenses involved in the import process of fuel makes the subsidy fund bogus. For instance, from the current petroleum product template outlined by PPPRA, it is clear where part of the funds go to. Beside the product cost which is the main cost, all other costs could be regarded as extra and avoidable, such as storage

Chapter Eight	Adadu Y. Abdullahi, Usman A. Tom & Hafsat Kigbu, in Jideofor Adibe (Ed.) *The Politics and Economics of Removing Subsidies on Petroleum Products in Nigeria* London & Abuja, Adonis & Abbey Publishers

185

charges for using depots, landing cost and distribution margins. This clearly shows where a large chunk of the subsidy fund goes to.

Conclusion and the Way Forward

This paper examined petroleum product pricing and the manipulations the subsidy in them in Nigeria. It argued that there is a serious problem with the line of supply of petroleum products due to the poor performance of the country's refineries which barely operates at about 15% of full capacity. The paper further argues that though fuel importation is relied upon as a way ameliorating the out supply shortage. It created an attendant problem of adjusting domestic prices in the attempt to offset the cost implication. The government fail to realize that price deregulation without efficient local supply base provides further incentives for fuel importation by the 'oil cabals' to consistently defraud the Federal Government by cooking up the product delivery account.

As stated earlier, the government itself has admitted in its recurring battle against subsidy, that the rich and the oil cabals, and not ordinary Nigerians are the beneficiaries. Eliminating the cabals with all the capacity its members have been using in manipulating the price of oil, will assist a great deal in bringing down the subsidy to a sustainable level.

There is no doubt the problems of fuel subsidy is not restricted to revenue loss but lack of transparency in the management of the subsidy fund as recently revealed by the House of Representatives fuel subsidy probe headed by Farouk Lawal. To this, the relevant Anti-corruption agencies should regularly carry out a diligence investigation on the payment to importers by NNPC and PPPRA.

Since the policy focus of government is the liberalization of the oil sector, this has to be done with caution if the poor are not to be adversely affected. It is advisable to ensure that the local refineries are

| Chapter Eight | Adadu Y. Abdullahi, Usman A. Tom & Hafsat Kigbu, in Jideofor Adibe (Ed.) *The Politics and Economics of Removing Subsidies on Petroleum Products in Nigeria* London & Abuja, Adonis & Abbey Publishers |

186

made functional to guarantee increased in domestic fuel supply before liberalization or de-subsidisation.

Finally, looking beyond de-subsidization, government's search for alternative sources of revenue need to be underscored and appreciated in the context of bridging the gap that exist between production and consumption. For the average poor, while they are not entirely opposed to subsidy removal, the pathway for a secured future lies on government's emphasis on cutting cost of governance. In a sense, strengthening the link that exists between the major drivers of growth in the area of agriculture and manufacturing must be overhauled, within an overall goal of diversifying the sources of revenue that has become characteristic of economies that rely almost solely on a principal source of revenue.

References

Ajidani, M.S. and Sani, J. (2005), 'Instability in fuel price in Nigeria: Its implication for the Nigerian Economy', in *proceedings of the Annual conference of the Association of Economic Education*, (Akwanga, College of Education).

Okonjo-Iweala, N (2012), *Brief on fuel subsidy*, (Abuja; Federal Ministry of Finance).

Egwaikhide, F.O et al. (2004) 'The Unending Fuel Crisis in Nigeria; in Garba, A.,Ekwaikhide , O.O, F and Adenikinju, (eds), *Leading Issues in Macroeconomic Management and Development*, (Ibadan, Nigerian Economic Society).

Gupta, S., Benedict, C., Kevin, F and Gabriela, I. (2002) 'Issues in Domestic Petroleum Pricing in Oil Producing Countries,' *IMF working paper 02/140*, (Washington, International Monetary Fund).

Mohammed, Y.A. (2011), *the parameters of political science* (Lafia, Rafa printing and publishing).

Oluleye, F.A. (2005), *Deregulation and Globalization in Nigeria: Issues and Perspectives.* (Ekpoma, AAU publishing house)

Chapter Eight | Adadu Y. Abdullahi, Usman A. Tom & Hafsat Kigbu, in Jideofor Adibe (Ed.)
The Politics and Economics of Removing Subsidies on Petroleum Products in Nigeria
London & Abuja, Adonis & Abbey Publishers

Roberts, F.O. (1998), *the politics of petroleum products pricing in Nigeria.* (Ibadan, NISER monograph series No 4).

Ukpong, G.E. (2004), 'Pricing of Petroleum Products in Nigeria and the Issue of Price Subsidy', *CBN Economic and Financial Review'* Vol.42, No 4, December.

Daily Trust (2011), 'The manipulation of oil subsidy', 2 November.

Daily Trust (2011), 'This "Subsidy" non-debate is in a cul-de-sac,' 20 December.

Daily Trust (2011), 'Subsidy: Gilding the Lily', 21 October.

Daily Trust,(2011) 23 December

West Africa Insight (2012), Vol.3 No 4, February.

This Day (2012), 2 January

The News magazine (2011), 9 December.

Chapter Eight | Adadu Y. Abdullahi, Usman A. Tom & Hafsat Kigbu, in
Jideofor Adibe (Ed.)
The Politics and Economics of Removing Subsidies on Petroleum Products in Nigeria
London & Abuja, Adonis & Abbey Publishers

Conclusion

De-subsidization of PMS: Beyond the debates and politics

Jideofor Adibe

Nearly two years (i.e. by October 2013), after the sharp reduction in the level of subsidies, it may be germane to take stock. What is the impact of the sharp-reduction in the level of subsidies on the economy and on social welfare? In particular what has been the impact of the drastic reduction in the level of subsidization of PMS against the arguments of the pro-desubsidization lobby?

Assessing the impact of the drastic reduction in the level of subsidizing fuel against these parameters obviously presents challenges, especially when that assessment inevitably has to be done using the 'before' and 'after' method. For one, many good measures in the country tend to suffer at the level of implementation, therefore even if the impacts of the drastic reduction of the subsidy on fuel on the economy and on social welfare point to the opposite direction from that claimed by the pro-desubsidization lobby, they can still blame it on implementation bottlenecks rather than the theoretical premises of their arguments. In essence, it will remain hypothetical what would have happened if the measures were implemented as they should or as they appeared on paper or in theory - without the inevitable interface with what is popularly known as the 'Nigerian factor'. The other problem is that there are other variables at play, including the impact of the continued challenge from insurgency groups like Boko Haram and other political and economic variables that could affect policy performance. Therefore the 'before' and 'after' methods of assessment cannot be a reliable way of assessing the impact of the de-subsidization against the arguments of the pro- and the anti-desubsidization lobbies. Despite these, nearly two years after

Conclusion | Jideofor Adibe (Ed.)
The Politics and Economics of Removing Subsidies on Petroleum Products in Nigeria
London & Abuja, Adonis & Abbey Publishers

189

the sharp reduction in subsidies, there are pointers that could, at least through anecdotal evidence, be attributed to the effect of the policy.

One of the issues that quickly emerged in the wake of the sharp reduction in the level of subsidy on January 1 2012 and the 'shut-down Nigeria' protests it triggered, was that the whole subsidy regime was tainted with corruption and was an avenue of illicit enrichment by a cabal of oil marketers.

In fact on January 8, 2012, the House of Representatives met in an emergency session and decided, among other things, to set up an ad hoc committee to probe the management of the fuel subsidy scheme following the uproar and protests that accompanied the government's ambush announcement on January 1 2012 that it had completely removed subsidy on fuel when it had given the impression that it was still consulting on the issue.

The House named an eight-member ad hoc committee led by a four-term member of the House, Farouk Lawan, to probe the subsidy regime between 2009 and 2011. The aim was to uncover any of the widely suspected corruption in the management of the subsidy regime and to unmask any cabal that had been feeding fat on the federal government's Petroleum Support Fund at the expense of the masses.

The Lawan committee conducted its probe in public, received memoranda from members of the public, and invited major stakeholders in the oil industry to testify before it. The hearings lasted for about three weeks and several stakeholders were invited, including "93 oil marketers and importers, the Nigerian Navy, the auditors appointed by the finance ministry to audit and verify subsidy claims, Federal Road Safety Corps, the professional bodies in the downstream oil sector, foreign oil traders, the Nigerian Labour Congress, Trade Union Congress, the managing directors of the Port Harcourt, Warri and Kaduna Refineries, Revenue Mobilisation, Allocation and Fiscal Commission, NIETI, and private individuals" (ThisDay, 15 July 2012).

Conclusion | Jideofor Adibe (Ed.)
The Politics and Economics of Removing Subsidies on Petroleum Products in Nigeria
London & Abuja, Adonis & Abbey Publishers

190

The committee took testimonies from 130 witnesses and received in evidence 3000 volumes of documents from January 16 to February 9. One of the most startling details that emerged from the Lawan committee's probe was how in 2011 the country paid subsidy on 59 million litres of petrol per day when in fact the daily consumption was 35 million litres (ibid). The Lawan committee submitted its report to the House on April 19, 2012 after sitting for three months. The House adopted the report on April 24 for onward transmission to the executive for implementation.

However, shortly after this, the Chairman of the oil subsidy ad hoc committee, Farouk Lawan, was accused of demanding and receiving bribes from one of the oil marketers, Femi Otedola, in what many people regarded as a fight-back by the indicted oil cabals (Adibe, 2012).

As the Lawan Report became mired in controversy, the Federal Ministry of Finance set up a technical committee in May 2012 to scrutinise the fuel subsidy payments to marketers during the 2011 financial year. The Ministry of Finance was miffed that despite paying N1.7 trillion in subsidy by December 2011 and another N450 billion in early 2012 to clear the 2011 backlog of debt owed to the oil marketers, outstanding claims of subsidy payment still remained from 2011.

The finance ministry's technical committee was headed by banker and member of the Economic Management Team (EMT), Mr. Aigboje Aig-Imoukhuede, and comprised representatives of the Central Bank of Nigeria (CBN), Budget Office of the Federation, Debt Management Office, PPPRA, Independent Petroleum Marketers Association of Nigeria (IPMAN), Major Marketers Association of Nigeria (MOMAN), and Office of the Accountant General of the Federation. In addition to its members, the technical committee also sought the support of CBN examiners and other financial experts and consultants, including Lloyd's List Intelligence, which was also consulted by the House of Representatives' ad hoc committee.

Whilst the House of Representatives' ad hoc committee comprised lawmakers, the Ministry of Finance's technical committee was made

Conclusion | Jideofor Adibe (Ed.)
The Politics and Economics of Removing Subsidies on Petroleum Products in Nigeria
London & Abuja, Adonis & Abbey Publishers

191

up of experts or professionals. The committee submitted its report to the Minister of Finance, Dr. Ngozi Okonjo-Iweala about six weeks after it was set up. Efforts were made to rubbish the work of the technical committee with the oil marketers and importers criticizing it for not giving them a chance to defend themselves, an allegation that prompted President Goodluck Jonathan to set up a presidential committee, also headed by Aig-Imoukhuede, to verify and reconcile the findings of the technical committee. The presidential committee was given about a week to submit its report.

As expected, the affected oil firms in the three reports (i.e. the House of Representatives' ad hoc committee, the Ministry of Finance's technical committee and the presidential committee) did not hesitate to fight back in order to 'clear their names'.

An early indication of how powerful the oil marketing cabals were came when the federal government announced that it would not prosecute the firms indicted by the House of Representatives' ad hoc committee, citing the allegations of corruption and extortion that trailed the Lawan report. At the same time several of the indicted firms, which had contested the veracity of the House's report, went to court to try to exonerate themselves alleging that they were not given fair hearing by the committee.

Media campaigns, which were suspected to be sponsored by the indicted oil cabals, were also waged against Aigboje Aig-Imoukhuede, the chairman of both the Ministry of Finance's technical committee and the presidential committee. For instance, Access Bank, which he heads, was accused of "financing the importation of about 40 per cent of the petrol brought into Nigeria in 2011 and received, into accounts at its branches, a similar percentage of the N2 trillion the FG paid as subsidy during the period" (Sahara Reporters, 2012). It was also claimed that Aig-Imoukhuede's bank actually recommended some of the indicted oil marketers. For instance, it was claimed that "Spog Petrochemicals which was alleged in a petition to the Attorney General of the Federation in 2011 to have imported 3000 metric tons of petrol despite receiving subsidy for 13,000 metric tonnes in early

Conclusion | Jideofor Adibe (Ed.)
The Politics and Economics of Removing Subsidies on Petroleum Products in Nigeria
London & Abuja, Adonis & Abbey Publishers

192

2011, came highly recommended by Mr. Aig-Imoukhuede's Access Bank" (ibid). It was equally alleged that Aig-Imoukhuede owned a company called Ice Energy and Petroleum Trading Company Limited with its office at Lake Chad Crescent, Maitama, Abuja. The company was alleged to have collected USD 2,131,166.32 (N345. 3 m) in 2011 without importing any petroleum products (Nigeriafilms.com, 2012).

What seemed to emerge from the investigations of the subsidy regime was that the oil marketers were perhaps far more powerful and connected than the general public initially thought. This in turn appeared to lead to some opponents of de-subsdization shifting to support the measure, but on provisos that such must be phased and that the people should be carried along (Agbedo & Akaan, 2012). But even among those who favour phased de-subsidization, there remained a lingering concern about whether any savings from such a measure would be managed transparently and efficiently. As John Iyobhebhe (2013) put it:

> ...The majority of Nigerians do not feel the impact of the oil wealth that is so talked about in Nigeria. The subsidy is an indirect form of wealth redistribution to the poor majority. If this is taken away then the government must come up with policies to compensate Nigerians, utilize the savings and explain how the inevitable inflation will be managed.

One of the arguments of the government in support of de-subsidization was that subsidizing fuel was making it difficult for it to meet its budgetary obligations. One would therefore have expected that with the reduction in the level of subsidy effective from January 1 2012, government would no longer have difficulties in meeting the budgetary obligations especially in the absence of shocks in the global oil prices or sharp reductions in the country's production. Contrary to this expectation however, there have been speculations that nearly two years after that move, when the federal government's finances ought to have become buoyed by the savings from the partial de-subsidization, the federal government might in fact have gone broke with state and local governments not receiving their statutory

Conclusion | Jideofor Adibe (Ed.)
The Politics and Economics of Removing Subsidies on Petroleum Products in Nigeria
London & Abuja, Adonis & Abbey Publishers

193

allocations as and when due for the first time since 1999. As Governor Matthew Oshimohole of Edo State put it:

"I don't know if the Federal Government is broke, but I know there is a serious financial crisis and it is unprecedented in the history of this country. That, for the first time since 1999, allocations can no longer come as and when due to states is shocking…. Two years ago, it was about the kind of money we were spending on subsidy. In no time, following series of probes and enquiries by the National Assembly and by the Presidency, they have since discovered the kind of money they stole as regards subsidy, all the people that conspired with them and I believe the EFCC is dealing with that. But just as we are dealing with that, now we begin to hear about the theft of our crude oil such that what is accruing to the Federation Account is not enough to meet budgetary provision" (Vanguard, 13 October 2013).

In essence, the government which claimed that unless it removed subsidies it would not be able to meet its financial obligations, in less than two years after its sharp reduction in the level of subsidy for fuel, became apparently enmeshed in worse financial crisis, and reportedly finding another excuse – alleged theft of crude oil – for its inability to meet budgetary provisions. Again for the first time since 2003, some civil servants at both federal and state levels were not paid their salaries as and when due. Obviously the fact that the sharp reduction in the level of subsidy for fuel has apparently not led to improvement in the federal government's finances as the pro-de-subsidization lobby claimed, will make it more difficult to convince the citizens of the necessity of another round of subsidization.

Another way of assessing the impact of the reduction in the level of subsidy is to critically examine the government's special intervention programme for managing the supposed savings from the reduction in the level of subsidy in January 2012.

Assessing the government's Sure-P programme

To assure Nigerians that the savings from the reduction of the subsidies on fuel would be managed transparently and to the benefit

Conclusion | Jideofor Adibe (Ed.)
The Politics and Economics of Removing Subsidies on Petroleum Products in Nigeria
London & Abuja, Adonis & Abbey Publishers

194

of ordinary Nigerians, the Federal Government on February 13 2012, constituted a 21-member Subsidy Re-investment and Empowerment Programme (Sure-P) Committee. The Committee's mandate was to manage and re-investment the federal government's share of the savings from the partial reduction of subsidies on petroleum products.

According to the Sure-P's website, the savings were to be "invested in programmes and initiatives that would bring relief to citizens experiencing any hardship from the subsidy withdrawal" (Sure-P, no date). The programme was also to ensure that the federal government's portion of the saving was "applied to critical infrastructure projects and social safety net projects that will directly and positively impact on the people" (ibid). Described as a special intervention programme, by the end of November 2012, it had according to the information on Sure-P's website spent the sum of N62, 423,351,736.58 on capital projects as summarized by the table below:

Table 1: Sure-P's Expenditure Profile, February 2012- November 2012.

S/No	Project	Project Implementation unit	Expenditure
1	Maternal Child Health	Federal Ministry of Health	N3,803,152,276.13
2	Public Works	FERMA	N4,000,000,000.00
3	Mass Transit	Infrastructure Bank	N8,900,000,000.00
4	East-West Road	Ministry of Niger Delta	N8,148,855,134.04
5	Roads & Bridges	Federal Ministry of Works	N28,296,238,063.10
6	Railway	Federal Ministry of Transport	N9,275,106,263.31
7	Secretariat Services	Sure-P	N325,525,292.27
Total			N62,748,877,038.85

Source: Sure-P website: http://www.sure-p.gov.ng/index.cfm/sure-p-the-journey-so-far/

Conclusion   Jideofor Adibe (Ed.)
The Politics and Economics of Removing Subsidies on Petroleum Products in Nigeria
London & Abuja, Adonis & Abbey Publishers

195

The above expenditures however do not tell us about the level of completion of the projects or the quality of work on them. As at the time of writing (October 2013), information on such was unavailable. Some critics have however argued that most of the projects numerated in the Table were already captured in the federal government's budget for 2012, meaning that the above were mere duplication of what the government had already earmarked funds for. This was also the position of Nigeria's National Assembly, which called for the scrapping of the programme (The Guardian, 28 November 2012). Others have criticised the whopping sum spent on staff salaries and emoluments - the so-called 'secretariat services' in the Table above - in just nine months (Vanguard, October 6 2013).

What is arguably the programme's most ambitious – and from poverty-reduction perspective most relevant - is its graduate internship scheme where it pays interns N10,000 per month while employers are allowed to match that – or pay more if they can. Though concerns had been expressed that such could lead to disguised unemployment and undue exploitation by employers, the idea of getting graduates off their beds, if for nothing else, to re-boost their confidence, appears commendable - though one could also legitimately question whether one needed to wait for de-subsidization for such a programme to kick off. However less than two years into the scheme, it would appear the programme has already run into difficult times. The Punch of September 20 2013 for instance quoted the chairman of the Subsidy Re-Investment and Empowerment Programme Committee, Dr. Christopher Kolade, as raising alarm that the intervention body would no longer be able to pay the N10,000 monthly stipends to each beneficiary of its scheme from the end of that month due to lack of funds. Dr Kolade lamented that the inability of his committee to honour its obligations to the 111,000 youths it engaged across the country was because the senate

Conclusion | Jideofor Adibe (Ed.)
The Politics and Economics of Removing Subsidies on Petroleum Products in Nigeria
London & Abuja, Adonis & Abbey Publishers

196

reduced its budget of N27bn meant for that purpose to N9bn. He was further quoted as saying:

> The actual amount needed to engage 5,000 youths across the 36 states and the FCT was N28.5bn but we presented a budget of N27bn which was reduced to N9bn. We immediately alerted the Federal Executive Council and we were assured that the original amount requested would be restored. However, since April when we were given the assurance, the amount was not restored hence we reduced the numbers of beneficiaries to 3, 000 per state but the fear now is that there is no more fund left in the budget and we won't be able to pay them at the end of the month if urgent steps were not taken to address the situation.

In addition to the above, which raises questions about the sustainability of the Sure-P programme, there were also several allegations of corruption and lack of transparency in the management of the programme both at the federal and state levels. For instance early in 2013, the Plateau State House of Assembly began the probe of over N5 billion suspected to be missing from the state government's share of the Sure-P accruals following public outcries in the state that the funds might have been misappropriated (Leadership, 19 August 2013). When the probe started, the ad hoc committee raised concerns about the refusal of the officials of the State government who appeared before it to give details of the bank accounts of the SURE-P funds. There was equally an allegation that the state government had been receiving N218 million monthly as its share of Sure-P funds, to an accumulated amount of N3billion at the time of the probe while it had also been receiving N146 million monthly on behalf of its 17 local governments to an accumulated sum of N2 billion (Vanguard, 6 October 2013).

In the same vein, in Kaduna State, the State Executive and the House of Assembly were also pitched in a war of attrition after the State's legislature earlier in the year took the state government to task over the composition and implementation of SURE-P in the state. The Chairman of the House's Ad-hoc committee on investigation into the

Conclusion | Jideofor Adibe (Ed.)
The Politics and Economics of Removing Subsidies on Petroleum Products in Nigeria
London & Abuja, Adonis & Abbey Publishers

197

implementation of SURE-P programmes, projects and activities in the state, Kentiok Irimiya Ishaku, reportedly said the actual receipts by the Kaduna State Government of its share of revenue from subsidy reduction for 2012 stood at N2, 243,188,906.24 at a monthly rate of N280, 398,613.80 from May to December, 2012. He claimed that there were no documented receipts for the months of January to April, 2012. The House accused the State's Sure-P Committee of stealing N560m from what accrued to the state from the programme (Channels, 27 June 2013).

The above instances show that concerns about transparency in the management of the savings from the drastic reduction in the level of subsidies were legitimate. This will obviously affect further moves by the government to remove the remaining subsidy as it will be seen by many Nigerians as transferring the opportunity for rent-seeking and primitive accumulation from oil marketing cabals to a government mafia and their cronies.

In essence, a general perception of the way the Sure-P has been managed, including its successes and failures, will affect people's willingness to accept any government rationale for removing the remaining subsidy – despite an apparent shift in sentiments that de-subsidization would perhaps be the most effective way of taking the fight to the door-steps of the oil cabals. For as long as people cannot be convinced that the savings will be managed well and transparently there will continue to be opposition to any further de-subsidization.

In addition to ensuring transparency in the management of Sure-P and that the programme makes meaningful impact in the provision of infrastructure and poverty reduction in a sustainable manner, the government also needs to provide the necessary frame work for the resuscitation of the country's refineries. There are estimates that the country's refineries, if resuscitated and developed further, could have a refining capacity of 385,000 barrels per day (Bpd) (Adekunle, 2013). Refining a substantial part of the country's domestic fuel consumption needs locally, is expected to help lower the unit price of fuel – even after full de-subsidization. In a country where trust in

Conclusion | Jideofor Adibe (Ed.)
The Politics and Economics of Removing Subsidies on Petroleum Products in Nigeria
London & Abuja, Adonis & Abbey Publishers

198

government is very low, conspiracy theorists believe that the non-resuscitation and modernization of the country four refineries, is a deliberate ploy by the government to ensure that huge quantities of fuel needed for domestic consumption are imported because top government functionaries are supposedly also profiting from the fraud in the subsidy regime.

There are several other recommendations both from contributors to the volume and in the course of the debate on whether subsidies would be removed or retained prior to the government's ambush announcement of January 1 2012. These include the need for the government to develop a comprehensive and viable power and energy policy, which will encompass the use of all fossil fuel and which will petroleum products, coal and gas.

The federal government has also been called upon to ensure greater transparency in the operations of the Nigerian National Petroleum Corporation (NNPC), Nigeria's primary representation in the oil industry. Currently the NNPC is generally regarded as being unwieldy and a cesspool of corruption, with its operations shrouded in so much secrecy. Nigerians do not even know the exact amount that has been saved from the reduction in the level of subsidy in January 2012. Many have called for the NNPC to be broken into a number of independent agencies and companies, and made to be more transparent in their operations.

Essentially a public perception of the way the Sure-P has been managed will affect the public's willingness to accept further or even complete de-subsidization. Though less than two years may not be enough time to pass judgment on whether the programme is succeeding or not, the tentative conclusion is that if the current trends are anything to go by, the omens are not good for the programme.

Conclusion | Jideofor Adibe (Ed.)
The Politics and Economics of Removing Subsidies on Petroleum Products in Nigeria
London & Abuja, Adonis & Abbey Publishers

References

Adibe, Jideofor (2012), 'Why we must rally behind Farouk Lawan', *Daily Trust*, June 21, back page.

Adekunle, oseni Semiu (2013): 'The impact of subsidy removal on Nigerian economy: series two' The National Economic Transformation, blog, 18 March,http://nationaleconomictransform ation.blogspot.com/2013/03/the-impact-of-subsidy-removal-on_340.html (Accessed 10 October 2013)

Agbedo, C.U. and S.S. Akaan, (2012): 'Fuel Subsidy Removal and Mind Control Game In Nigeria: A Critical Discourse Analysis Perspective', *Journal of Humanities and Social Science* (JHSS), Volume 6, Issue 2, http://www.iosrjournals.org/iosr-jhss/papers/V ol6-issue2/B0620616.pdf(Accessed 10 October 2013).

Channels, (2013): 'Kaduna Assembly Accuses SURE-P Committee of Stealing N560 Million', 27 June, http://www.channelstv.com/home /2013/06/27/kaduna-assembly-accuses-sure-p-committee-of-stealing-n560-million/ (Accessed 13 October 2013).

Iyobhebhe, John (2013): Removal Of Fuel Subsidy In Nigeria- The Issues & Challenges, *Nigeria Politico*, October 10, http://www.niger iapolitico.com/subsidy%20.html (Accessed 10 October 2013)

Nigeriafilms.com, (2012): 'Oil Subsidy Cabal's Plot to Rubbish Aig-Imoukhuede's Report Uncovered!! Jonathan Directs EFCC to Arrest Subsidy Thieves', July 15 http://www.nigeriafilms.com/ne ws/18200/4/oil-subsidy-cabals-plot-to-rubbish-aig-imoukhuedes.html (Accessed 9 October 2012).

Sahara Reporters, (2012): 'Access Bank, CBN, Others Culpable In $6.8 Billion Petrol Subsidy Fraud-PREMIUM TIMES', July 18, http://sa harareporters.com/news-page/access-bank-cbn-others-culpable-68-billion-petrol-subsidy-fraud-premium-times (Accessed 9 October 2012).

The Guardian, (2012): 'Senate panel wants SURE-P scrapped, second Niger Bridge to cost N7 billion', 28 November, http://www.ngrgu ardiannews.com/index.php?option=com_content&id=106167:senat

Conclusion | Jideofor Adibe (Ed.)
The Politics and Economics of Removing Subsidies on Petroleum Products in Nigeria
London & Abuja, Adonis & Abbey Publishers

e-panel-wants-sure-p-scrappedsecond-niger-bridge-to-cost-n7-billion (Accessed 13 October 2013)

ThisDay (2012): 'Petroleum Minister Lists Gains of Subsidy Removal' 10 January, http://www.thisdaylive.com/articles/petroleumminister-lists-gains-of-subsidy-removal/106795/ (Accessed 11 October 2013).

Vanguard, (2013): 'Sure-P: Sure pit for Nigeria's fiscal drain pipes, by Lai Mohammed' 6 October,http://www.vanguardngr.com/2013/10/sure-p-sure-pit-nigerias-fiscal-drain-pipes-lai-mohammed/ (Accessed 12 October 2013)

Vanguard, (2013): 'Nigeria in deep financial crisis – Oshiomhole', 13 October, http://www.vanguardngr.com/2013/10/nigeria-deep-financial-crisis-governor-oshiomhole/ (Accessed 13 October 2013

Conclusion | Jideofor Adibe (Ed.)
The Politics and Economics of Removing Subsidies on Petroleum Products in Nigeria
London & Abuja, Adonis & Abbey Publishers

Index

A

B

C

D

E

F

G

Index | Jideofor Adibe (Ed.)
The Politics and Economics of Removing Subsidies on Petroleum Products in Nigeria
London & Abuja, Adonis & Abbey Publishers

Index | Jideofor Adibe (Ed.)
The Politics and Economics of Removing Subsidies on Petroleum Products in Nigeria
London & Abuja, Adonis & Abbey Publishers

203

Index | Jideofor Adibe (Ed.)
The Politics and Economics of Removing Subsidies on Petroleum Products in Nigeria
London & Abuja, Adonis & Abbey Publishers

204

Trade Union Congress, 34, 41

Tukur, Bamanga, 20

U

United Arab Emirates, 17

United Nations Human Development
Index, 95

United Nations Millennium
Development Goals, 49

United States, Vi, Vii, 16, 126, 127, 131

Urban Mass Transit Scheme, 42

US Farm Bills, 16

V

Very High Fuel Subsidies Countries,
126

Vietnam, 132

W

West, David, 15, 40

World Bank, Vi, Vii, 13, 14, 26, 42, 50,
141, 145, 146, 147, 151, 154, 155,
156, 170, 180, 183

Y

Yahaya, Abdullahi Adadu, V, 24, 173

Yar'Adua, Umaru Musa, 15, 74, 179

Z

Zambia, 128

Index   Jideofor Adibe (Ed.)
The Politics and Economics of Removing Subsidies on Petroleum Products in Nigeria
London & Abuja, Adonis & Abbey Publishers

205

www.ingramcontent.com/pod-product-compliance
Lightning Source LLC
Chambersburg PA
CBHW030650270326
41929CB00007B/296